RACING PIGEONS

Colin Osman's grandfather is regarded as the founder of the sport of pigeon racing in Great Britain. He began writing about pigeons in the 1880s and founded *The Racing Pigeon Weekly*, the management of which has been handed on through the generations of the Osman family. Apart from the highly successful *Racing Pigeons*, Colin Osman has written several other specialized books on the subject and is the author of many articles on pigeon racing.

T0174478

Racing Pigeons
A Practical Guide to the Sport

Colin Osman

faber and faber
LONDON · BOSTON

First published 1957
by Faber and Faber Limited
3 Queen Square London WC1N 3AU
Second Edition 1980
This new paperback edition published in 1996

Printed and bound by Antony Rowe Ltd, Eastbourne

A CIP record for this book
is available from the British Library

ISBN 0–571–17814–6

Contents

Illustrations

Introduction

The origins of this book go back to 1957 when I decided that although there were a number of specialist books on pigeon racing there was nothing that could be recommended as a basic guide for beginners. It is a sign of the times that in those days we all thought of novices as young and the book was written with the under-21s in mind. Today many novices are in their fifties and sixties. Early retirement and redundancy have meant that many men find that in their mature years they have time on their hands and a little bit of money they can spare for a hobby. The advantages of pigeon racing as that hobby are enormous. It is time-consuming. No one can hope to be successful if he is not willing to spend some time with his pigeons. Make no mistake, at the height of the season a man in work will have to spend at least one or two hours every day in the loft. If he can get home at mid-day he will spend more time on the pigeons than on his lunch. If the man has a wife who is at home at midday then he may well find that she can give vital help during the day.

In the old days, before my time, I suspect that many a man found peace and quiet with his pigeons away from the woman's world of the kitchen and scullery. Today he may still find his pigeons provide a release from everyday pressures, but he is far more likely to share his hobby with his wife. Although there are no regulations against it, few women race on their own but a large number race in partnership with their husbands.* Partnerships are nothing new. In the days when many fanciers worked in the coal mines a partnership would often be formed of men on different shifts so that not only could the workload be shared but it could also be spread over the whole day.

The advantages of racing pigeons for someone who has taken, or been given, early retirement are obvious. Not only does he

---

*For convenience, and to avoid constant use of 'he or she', I have therefore addressed this book at what I presume will be a largely male readership. I apologise if this causes any dismay to female readers.

have an occupation to fill his days but he also has access to a social life through the weekly club meetings and the annual dinner, dance and prize presentations of his own and other local clubs. There is also the possibility of making money! The novice should not be misled by the stories in the national press about the tens of thousands of pounds paid out for champion racers. In some cases the prices quoted may be genuine but often, regrettably, they should be regarded as payments for advertising. In my view they do not help the sport but only attract the wrong sort of person who thinks this is a way to get rich quick. That sort of person will soon be disillusioned and leave the sport as fast as he joined it. The average successful fancier will think that he has done well if his winnings over the year cover his costs. That is not difficult to do and there are a few who, at the end of a year, may even show a decent profit. In almost every case, these fanciers are the ones who have been at the top or near it for years, and then suddenly everything goes right and they finish up winning the National. Every fancier hopes to hit the jackpot but most know that this is not the real prize to be won.

Going back to the old days again, the racing pigeon was often called the poor man's racehorse. Until I studied racehorse breeding a little, I used to resent that title. Photography has always been a second hobby of mine and for some years, I freelanced as the official photographer at two greyhound racing tracks. My job was to photograph the winning dogs and their owners. It dawned on me that the owners probably only saw their dogs two or three times a year and that for the rest of the year the dogs were in kennels with dozens of other dogs controlled by trainers, handlers and kennelmaids. In the same way, racehorse owners were with their expensive animals to lead them into the paddock but rarely saw them apart from that. The racing pigeon owner is not only owner but trainer, breeder and even jockey as well as stable lad. Far from being poor he is rich indeed, for in the immortal words made famous by Frank Sinatra, when he wins, he does it his way.

Pigeon racing is a practical sport. It is not a sport that can be learnt from books alone but is something that must also be learnt with pigeons in a pigeon loft. This book does not offer an open door to immediate success, but is more an attempt to prevent the reader from making mistakes which could prove costly.

The sport's 150-year history has shown that in the management of pigeons some things are unquestionably right and others are just as wrong. On these matters good pigeon fanciers throughout the world are agreed. On other matters, however, there is far less agreement and in many cases no agreement at all. In this book will be found advice based on the general practice of successful pigeon racers and, where these disagree among themselves, then that also will be noted. It is a book of experience; not just the limited experience of the author but the experience of people with whom the author came into contact in his day-by-day work on *The Racing Pigeon*, a weekly paper read throughout the world.

When the monthly *Racing Pigeon Pictorial* was launched some twenty-five years ago to complement the weekly *Racing Pigeon*, there began a new era of writing about the sport. For the first time, colour was readily available, in particular for pictures of pigeons' eyes, and it was possible to take a longer look at the work of many of the outstanding pigeon fanciers. In this way, the monthly *Pictorial* supplemented what the weekly was already doing in bringing to pigeon fanciers the experience of the whole world. This experience has been analysed carefully, and some attempt has been made at deciding whether certain fanciers win because of the methods they adopt or in spite of them!

My job as editor was coupled with that of managing director, which meant that much of my time was taken up with printing contracts and accountancy. Others got the fun part. Now I have handed over the boring bits to my son, Rick, and being officially retired have more time to talk to other fanciers. I still do a little bit here and there, particularly with *Squills Year Book* and my visits to Belgium for the charity auction, but this is the fun part, both more enjoyable and more informative.

It is impossible in a book of this length to deal with every topic. As long as there are pigeon fanciers there will be fresh things to discuss and ten books could not cover all that could be written. Here in this book will be found the essential facts that a newcomer to the sport will need to know if he is not to be disappointed in his attempts to race pigeons.

Not everyone who takes up the sport will succeed. Indeed not all are suited by their temperament to enjoy the sport. Unless you are able and willing to study your birds as individuals and have the patience to wait, not just one but several years, to see the fruits of your labours, then you may well be advised to take up some less exacting hobby. If you are able to meet these requirements, you will find that keeping and racing pigeons can be the most profitable hobby you have ever had, not so much because it is possible to win prize money, but because by being guardian, teacher and judge of the birds, you will learn to love them and will gain something from them that is priceless.

ONE

The Pigeon Itself

The Skeleton – The Legs – Ringing – The Shoulder and Wing –
Wing Movement – The Blood Circulation – Respiration – Digestion –
The Cleansing of the Blood – The Endocrine Glands – The Feathers –
Colours and Patterns – The Rarer Colours – The Head and Eye –
Eyesign – Homing Ability – Sun-Arc Navigation – The ESP
Hypothesis

The racing pigeon is only one of the names that can be given to the subject of this book. It is also called the homing pigeon, the racing homer and even the carrier pigeon, although this last term is used only by those who do not realize that although the racer and the carrier may have been similar years ago, today the carrier is exclusively a show variety of pigeon of which only a very small number exists. Since the days of Charles Darwin, an early breeder of pigeons, the racing pigeon has been considered to be a close relative of the wild rock pigeon and indeed the same Latin name, *Columba livia*, is used by ornithologists for both. This is in some ways curious, since the wild rock pigeons are non-migratory and range only comparatively short distances from their homes. They would have no great use for the homing instinct which forms such a distinctive characteristic of the racing pigeon.

Experiments at Cambridge University have shown that the pigeon is not the only good homing bird. On one occasion a Manx Shearwater flew 3,050 miles (4,908 km) from America in 12½ days, the whole of the distance over water. There are hundreds of other examples, ranging from cats and dogs finding their way home from miles away to swifts returning to the same nest year after year. Migratory birds are not exactly the same because some only return to the natal area rather than to their exact birthplace. In this chapter we shall consider both the pigeon's body, which enables it to fly and race, and also what is known of its homing instinct.

The Skeleton

The central basis of the skeleton is the backbone. It is difficult to count the exact number of vertebrae in this since in the central portion of the back they are joined together as an immovable section. In fact, with the exception of the neck and the tail, the backbone is solid.

In the tail the bones are still separate with the exception of the final tail bone, the pygostyle. This bone is the base into which the tail feathers are fixed. The ribs are fixed to the top part of the solid section. These curve round the internal organs of a pigeon to join on to the breastbone. There are in addition two smaller ribs which are not fixed to the breastbone. They lie in front of the fixed ribs and also serve to protect the internal organs.

The Legs

The rear section of the solid part of the backbone is joined to other bones of the pelvic group including the pubic or vent bones. The two vent bones widen in the hen pigeon to allow the eggs to be laid. To this pelvic group are joined the leg bones. The thigh bone is normally concealed among the feathers on the side of the bird and this is joined to the shank, the 'drum stick' in a table fowl. On these bones the leg muscles are quite strong since to begin flight the pigeon jumps into the air. The bones are very similar in arrangement to the human leg except that the pigeon's ankle bone is rather like the elongated hock of a horse. Finally there are four toes having from three to five joints in each, with the exception of the small backward toe. This back toe has only two joints and is sufficiently flexible in the early days of a pigeon to allow the bird to have its permanent identity ring put on.

Ringing

To put the identity ring on the bird it is not essential to lift the squeaker out of the nest. The bird should be turned in its nestbowl until its right leg is accessible. As the bird should be about

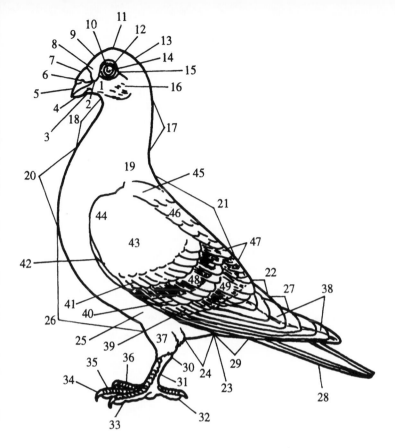

1 EXTERNAL ANATOMY.

Head: (1) Cheek, (2) Angle of lips, (3) Chin, (4) Lower beak, (5) Upper beak, (6) Nostril, (7) Cere or wattle, (8) Lores, (9) Frontal or forehead, (10) Pupil, (11) Crown or topskull, (12) Fringe of eyelid, (13) Backskull, (14) Iris, (15) Eye cere or orbital skin, (16) Ear feathers or auriculars. *Neck*: (17) Nape, (18) Throat or bib, (19) Neck blend or hackle. *Body*: (20) Crop, (21) Back, (22) Rump, (23) Vent, (24) Abdomen, (25) Flank, (26) Breast. *Tail*: (27) Upper or dorsal tail coverts, (28) Retrices or main tail feathers, (29) Under or ventral tail coverts or fluff. *Foot*: (30) Hock, (31) Tarsus or shank, (32) Hind or first toe (hallux), (33) Outer or fourth toe, (34) Claw, (35) Middle or third toe, (36) Inner or second toe, (37) Leg or tibia. *Wing*: (38) Primary remiges or wing flights, (39) Secondary remiges or wing flights, (40) Secondary or greater wing coverts, (41) Median or middle coverts, (42) Bastard wing of thumb feathers, (43) Lesser coverts, (44) Wrist or wing butt, (45) Shoulder, (46) Scapulars or 'heart' or saddle, (47) Tertials, (48) Second wing bar, (49) First wing bar.

3

6–10 days old it will not actively resist handling but great care should always be used to make sure that the bird is not unnecessarily frightened or hurt during the ringing process otherwise the bird may become timid and unwilling to become friendly with the fancier. The ring is held upside down and slipped over the first three toes. The fourth toe will bend backwards against the shank as the ring is slid up the leg. The fourth toe must then be eased clear of the ring.

The idea of putting on the ring upside down is that when the bird is held correctly in the right hand with its head forward and with the thumb across the back and with the bird's feet between the first and second fingers, then the left hand can be used to extend the bird's foot. The ring will then be in an easy position to

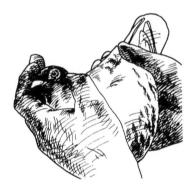

2 HOW TO HOLD A PIGEON SAFELY.

read. It is all very simple, but because some fanciers forget this, endless time can be wasted when the birds are being marked for races.

This is the usual procedure but I am a bit of a rebel and ring on the left leg, the bird facing away from me. I argue that this means that when the bird is race marked the leg in your right hand is more convenient to read, and after the race the rubber ring is easier to remove from the other leg. It doesn't matter much and I do have to say I seem to be in a minority of one.

The Shoulder and Wing

A group of three bones form the shoulder of a pigeon, the short coracoid, the clavicle and the scapula. The scapula or shoulder blade moves over on the backbone. The coracoid joins on to this bone at the wing butt and pivots on the breastbone. The clavicles from each wing butt join together and form the wishbone of the pigeon. This complex arrangement of bones has a most important part to play in the movements of flight. A fourth bone joins on to the shoulder at the wing butt, the humerus, the upper arm bone of a pigeon.

The bones of the wing of a pigeon are similar in many respects to the arm of a man; the lower arm has two bones like a human forearm. After the wrist joint the bones, as in the case of a pigeon's foot, are considerably elongated compared to human bones. They are joined on to the finger joints or the phalanges of the wing. Also at the wrist joint is the pollex or thumb. Round this thumb is

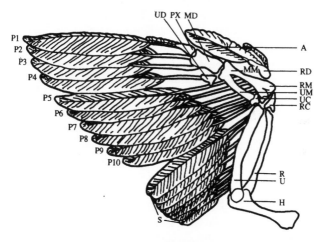

3 THE SHOULDER AND WING.
P1–P10, primaries; S, secondaries; H, humerus; R, radius; U, ulna; RC, radial carpal; RM, radial metacarpal; UC, ulna carpal; UM, ulna metacarpal; MM, middle metacarpal; MD, middle digit; UD, ulna digit; RD, radial digit; PX, phalanx; A, alula or bastard wing.

a tuft of feathers known as the bastard wing. The exact purpose of this bastard wing is not known, but it is suggested that it is used by pigeons on landing to break up the air flow over the wing and acts very similarly to the way slots are used on some aircraft.

Wing Movement

The movement of a pigeon's wing in flight has been greatly misunderstood and it was not until Francis Jenkins had made his high-speed film of the flight of birds and until electronic flash had made possible photographs of pigeons in flight with exposures of one-thousandth of a second that detailed study became really possible. The downward stroke of a wing consists of one sweep from right above the shoulder to right underneath. The wings actually touch above and below when a bird is taking off. The bones remain in the same relative position throughout the whole of this downward movement and each feather on the wing is held hard against the next one by the pressure of the air which is being displaced by the movement of the wing. Nearly the whole of the power of this downward stroke is the result of one muscle, the greater pectoral. This pair of muscles alone makes up nearly one-third of the weight of a pigeon.

The upward movement of the wing is much more complicated. The wing folds at the elbow joint. This folding is done by the biceps. At the same time the upper arm is lifted until the wing butts are almost meeting over the pigeon's back. At the end of this first stage of the upward stroke of the wing, the upper arm is vertically above the shoulder and the lower arm is still extended horizontally. In the second stage of the upward movement, and the slow-motion film shows these quite clearly to be separate stages, the wing is straightened by the triceps muscle. During the whole of the movement, but more particularly during the second stage, the feathers are actually rotated by some of the small muscles inside the wing, so as to allow the the air to pass between them. The action of the feathers is in a sense like a Venetian blind, allowing air to pass through them on the upward stroke and presenting an unbroken surface to the air on the downward stroke.

It was thought for many years that this rotation of feathers was the principal way in which the upward stroke was made easy, but the folding of the wing at the butts during the upward stroke is probably of greater importance. In addition there is yet another way by which a pigeon is able to save energy on the upward stroke, so that it can be used more usefully on the downward stroke. When the construction of a feather is considered in detail later, it will be seen that the web consists of barbules and barbicels. These barbicels hook together to provide an unbroken surface on the downward stroke, but unhook automatically on the upward stroke.

The Blood Circulation

The healing power of pigeons has often been noted and commented upon. A bird will return from a race badly injured after hitting an unmarked wire, but within a matter of days the wound heals up and the bird, after being rested, can be sent again. One fancier in one of my clubs won a race with a bird that only two weeks previously had had two large gashes in its crop sewn up. This healing ability is due to the rapid circulation and the richness of the blood. The heartbeats of a pigeon are much faster than those of a human heart. The pigeon is pumping round blood rich in both red and white corpuscles. The special function of the white corpuscles is the healing of damaged tissue and the combating of disease; thus the high rate of blood circulation makes possible the rapid healing of injuries. The blood has, of course, the other functions of carrying digested food throughout the body and of transferring the life-giving oxygen to every part of the body.

Respiration

Air is breathed in through the beak and nostrils, passes through the cleft in the palate of the bird down to the lungs. Here the oxygen is absorbed into the bloodstream in the walls of the lungs. This blood, rich in oxygen, is pumped through the arteries. The arteries break down into smaller and smaller tubes until the finest of all the capilliaries are reached and the oxygen is transferred out of the blood to give its life-giving energy to the whole of the body. As the blood loses its oxygen it changes from the bright red of the arteries to the bluish purple of the veins, and continues its journey back to the heart to be pumped round to the lungs again and be recharged with oxygen.

The action of the lungs in the pigeon differs from that of human lungs. In the bird the lungs are small and the air, when it is breathed in, passes through the lungs into the air-sacs. These

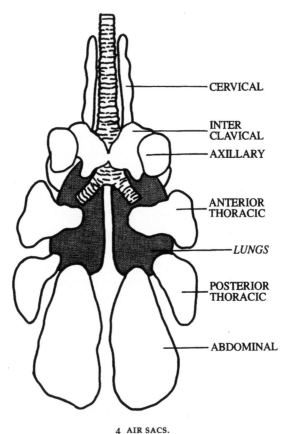

CERVICAL

INTER
CLAVICAL

AXILLARY

ANTERIOR
THORACIC

LUNGS

POSTERIOR
THORACIC

ABDOMINAL

4 AIR SACS.

8

air-sacs, nine in number, form cushions for the internal organs inside the ribs, except for the one in the throat near the crop. They enable the lungs to use clean air both on breathing in and breathing out and, in addition to this double action, are the equivalent of sweating in a pigeon. The pigeon has no pores in its skin for sweating and so excess water vapour is absorbed into the air in the air-sacs and expelled by the action of lifting the breast-bone. It should be noted that the active part of breathing in a pigeon is breathing out, not breathing in, as in the case of human beings.

Digestion

The blood is also vital to the digestive system of a pigeon. The simplified diagram helps to show the complicated processes by which the bird is able to use the food it is given. The food is picked up by the beak between the upper and lower mandibles. It is then passed into the gullet and then into the crop. In the crop the process of digestion begins, the peas or beans that the bird has eaten are softened and they pass from there into the glandular stomach. In the stomach the liquids from the glands are emptied over the food and the process of breaking it down into its basic

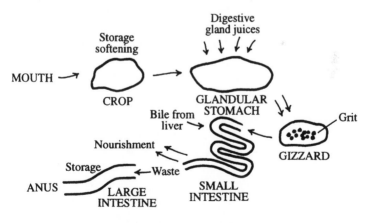

5 THE DIGESTIVE SYSTEM.

elements is continued. From the glandular stomach the food will move into the gizzard. This is a vitally important organ for a pigeon, for it replaces the function of teeth; it is here that the food is ground up into pieces small enough to be completely digested. This grinding process is accomplished by grit. Wild pigeons get this from the fields or rocks, but domestic birds must have it given to them. It can easily be understood that the birds' digestion and health will be endangered if insufficient or dirty grit is given to them. The grit that a bird eats can stay in its gizzard for weeks, even months, so there is never any cause for alarm if grit is not available for a day or two.

Grit is normally bought in bags from a supplier. Each bag contains hard grit that is often broken flints. This is not absorbed by the pigeon but as it grinds the corn so it gets worn away until it is so small that it passes through the system. This can take weeks or months, possibly years. The so-called soft grit is not really soft at all but is limestone rock that also grinds but becomes more rounded, and so has a crushing action as well. It does not last as long as the flint grit because the digestive juices work on it and eventually it is absorbed into the bloodstream, but even with this we are talking about days and weeks rather than hours. Many pigeons will seem anxious to get at the grit when a little pot of it is put in the loft but usually this depends on what extra has been added to the grit. It is quite usual to add minerals but at least one commercial brand has a lot of salt that the pigeons go for. A few fanciers make their own grit from sea shells gathered on the beach but obviously not many can do this. It seems to work although if I were using it I would want to add some crushed flints. It is a lot of work to save the price of a bag of grit.

Baby pigeons in the nest get their grit only when it is regurgitated with the food from the parents. That is why I make sure that when birds are rearing there is plenty of grit before them all the time. Sometimes I don't think the system works all that well and I am never happy to see 6- or 8-day-old chicks with their crops bulging with round beans or peas. They do get digested but it seems very inefficient and this is why for these critical days I like to feed pellets. The parents pump these into the youngsters and their baby crops are full and smooth with a food of equal goodness but much more digestible. In the chapter 'Food and Feeding' I will talk more about the nutritional benefits and recent developments in the use of pellets.

It is in the small intestine that the blood begins to play its vital part. The process of digestion is completed here, and the proteins, carbohydrates, fats, sugars and starches of the food are converted into a form of sugar which can be absorbed into the bloodstream through the walls of the small intestine. Once this food is in the bloodstream it is pumped round, together with the oxygen from the lungs, to feed all parts of the body.

The Cleansing of the Blood

The work of the blood is not finished even here, for in addition to the waste matter of indigestible food which is temporarily stored in the large intestine and then excreted, waste products are formed in the internal organs of the bird itself. The white corpuscles, which destroy disease, are themselves destroyed by their healing action. These and other waste matter must be filtered from the bloodstream; this is a job of the kidneys. The liver is also a blood filter but its main job is to make bile. The bile is poured into the small intestine to help in the digestion of proteins and fats. Needless to say, both these organs must function perfectly if the bird is to be in perfect health. The efficient functioning of the kidneys can be seen in the white toppings to the main dark mass of a pigeon's droppings.

The Endocrine Glands

The endocrine glands are those that act on the body through the bloodstream releasing hormones that stimulate other parts of the body. There are several of these glands but the most important of these is the pituitary since this one releases hormones which control the other glands as well. It is the master gland. The pituitary also produces the hormone prolactin which is responsible for the formation of the crop secretion known as 'pigeon milk'.

The thyroid glands lie near the throat of the pigeon and their hormone is necessary if a bird is to have good feathering and good bone structure. Iodine is necessary for this gland to work efficiently. Close to the thyroid glands lie the parathyroids. These are necessary so that the pigeon can use calcium and phosphorus. Ultraviolet light and vitamin D stimulate these.

The adrenal glands are situated just above the kidneys and these produce two entirely different hormones. One is adrenalin, which quickens the heartbeat under the stimulation of fear or excitement, and the other is cortisone, which is important both for its effects on the blood and on cell growth.

Some glands have an endocrine function as well as their more obvious one. The male and female sex organs, in addition to their normal function, secrete hormones which control the masculinity or femininity of the bird.

The pancreas, as well as making pancreatic juice, makes the hormone insulin which is vital for changing food into blood-sugar. In addition all of these glands tend to stimulate the others, so a correct balance is necessary for the efficient functioning of the bird.

This interaction of these glands is most important. The pituitary is the master gland, rather like the conductor of an orchestra, but each gland reacts to the others to a greater or lesser extent. When we come to the chapter on 'Winning Systems' the importance of this will be demonstrated.

The Feathers

The pigeon both flies and is kept warm by its feathers. Some are more adapted for keeping it warm while others are more useful for flight. The two sorts differ not only in size, for the flight feathers are larger, but also in construction since in flight feathers each little part of the feather locks together while non-flight feathers are fluffy to varying degrees. They all begin on the tiny naked youngster as down. This down is a lot of tiny clumps of yellow hairs. Behind these tiny tufts, the feathers are formed in sheaths. These feathers emerge gradually from their sheaths until the bird is completely covered at about five weeks old.

The flight feathers include the primaries on the outer part of the wing, the secondaries on the inner part of the wing and the retrices or tail feathers. These feathers although differing in shape are very similar in basic structure. Each is, in fact, immensely complicated and intricately detailed. Down the centre of each feather runs the quill, a hollow U-shaped tube of considerable strength. This fits into a socket of the wing and is held by the internal muscles of the wing. From each quill the barbs branch out on either side. There are well over 1,000 of these barbs on a normal primary feather. Each barb in turn has, branching out on either side of it, tiny barbules. These barbules when examined under a microscope will be seen to be of two types, one type having hooked barbicels and one which does not have these hooks. As there are a number of barbicels on a barbule, 1,000 barbules on a barb and 1,000 barbs on the quill, it can be seen that each single feather consists of many millions of parts.

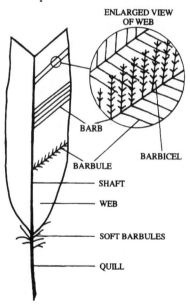

ENLARGED VIEW
OF WEB

BARB

BARBICEL

BARBULE

SHAFT

WEB

SOFT BARBULES

QUILL

6 THE STRUCTURE OF A FEATHER.

The barbicels, those with the hooks and those without, are arranged so that when the feather is pressing down on the air, they lock into each other, forming an air-tight surface. When they are moving in the reverse direction, that is upwards, they automatically unhook and allow the barbules to separate and the air to pass through. In this way not only do the feathers of the wing act like a Venetian blind, presenting a continuous surface on the downward stroke and open gaps on the upward stroke, but also inside each individual feather this same process is going on. It is this, of course, which enables a pigeon to save its energy on the upward stroke of the wing, so that it may use it more effectively on the downward or working stroke.

If a bird is held in the normal position with its head towards one's body and the tail pointing away, the wing can be fanned quite gently and the primary flights examined. It should be noted that particular stress has been laid on the primary flights rather than the secondaries, for although these latter undoubtedly do have a lifting action, their importance is completely overshadowed by the primaries. There is a Belgian theory which attaches a great importance to the secondaries but I have never seen practical demonstrations to support this theory, and it does not seem to be borne out by the adaptation of wild birds. For example, the bird with the most developed secondaries of all is the albatross and this bird is adapted particularly for slow effortless gliding flight and not for the high speed that we expect from the pigeon.

Colours and Patterns

The two main colours of racing pigeons are red and blue. The blue is very similar to that of its early ancestor, the rock dove. The body and wings are a grey-blue colour and across the wings there are two black stripes or bars. The head and neck is usually slightly darker and the bird may have a white patch on the rump just above the tail. This colouring is known quite simply as blue or blue bar.

In addition to this pattern the very common pattern of blue and black speckling is known as chequering. In this, little triangles of dark colour are found in a regular pattern over the wings. The bird can be lightly or heavily chequered depending upon how much grey-blue body colour is visible between the black chequers. When the wings are almost completely covered by the dark chequering the bird is usually called a velvet and when completely covered so that none of the lighter colour is visible at all, a black. Unfortunately the colours are very hard to describe in words and fanciers often disagree on the colours of birds.

The other main colour is a red-tinged grey with markings in a deeper red-grey colour. Chequering and bars appear in the red colour similarly to those of the blues, but the red bar is usually called a mealy. It is possible for mealies to be bred without bars although these barless mealies, like barless blues, are very rare. This normal red colour is called ash red by scientists to distinguish it from another and rarer form of red known to them as recessive red and known to most pigeon fanciers as brick red or sometimes chocolate. The different terms indicate how far this colour can vary but normally it will be found to be a much richer red colour than the ash red. Ash red primaries are always much lighter than those of recessive reds. To make things a little more difficult, there is a brown colour which is also referred to as chocolate, though dun is a better name since it is more a milk chocolate colour than the plain chocolate of the brick reds.

Grizzling is just one of the many ways in which white can appear in a pigeon's colour. It is recognized as an uneven speckled arrangement of white on the head and neck of the bird and also by the distinctive colour pattern on the flights. It is, in a few cases, difficult to distinguish from pied marking for pied birds also have white markings on the head, but normally these are smaller in number and larger in area and there is not the same distinctive colouring to the feathers. Pied is very frequently coupled with white flights, in which one or more of the primaries are white. Yet another form of white, very common in certain families, is the appearance of a very small patch of white just behind the eye; this is known as tic-eye.

Although there are a number of white pigeons, very few of these are albino white. Most of the white racing pigeons breed youngsters with patches of some other colour. These are usually called white pieds and when the patch of other colours appears on the back, as it does quite often, this is known as a red or blue saddle. A pied which is more than half white is usually called a gay pied. All these white patches can occur with all the usual colours of pigeons and with many of the more unusual ones.

The Rarer Colours

The red, blue and brown colours all have a dilute form. These are colours which are not very popular for racing and so do not turn up very often in lofts.

The dilute colours in all cases are a watered down form of their stronger or more intense counterparts. The dilute of red is known as yellow and is really a grey buff colouring. The dilute of blue and black is silver; this is the watered down form of blue in which the body is much greyer than in the intense form. The brown or dun colouring is thinned out to drab. As well as the markings being diluted, in many cases the body colour may be strengthened and so in the case of the silvers the flights, normally the ashy colour, appear slightly more yellowish. The drab colour is sometimes known as khaki and the silver chequering is sometimes referred to as lavender.

Dilutes are easily recognized at birth because they are always bald, that is they are born without the yellow hairs all other pigeons are born with. This is a simple example of genetic linkage for as the gene for dilute is transferred it brings along its close partner, the gene for baldness. Perhaps it is this seeming mystery that has given many fanciers the idea that dilutes make good stock birds. I have seen no scientific evidence for this idea but it may simply be that a fancier can separate the bird out at birth and give it a chance to prove itself at stock. He may also be influenced by the belief that dilutes do not make good racers because the feather changes that make them dilute also affect the barbicels in the feather, making the flights less waterproof. Some years ago I did some breeding experiments with dilutes and I came to the conclusion that they were less reliable in bad weather; however, not many pigeons were involved so it was hardly conclusive proof.

Among the rarities are some birds with triple colouring. Sometimes birds are bred with one red wing and one blue but these are genetic accidents, sports, that do not reproduce themselves. In a proper triple colouring there is a regular pattern of red and blue on a third body colour, and this can be bred from bird to bird. Bronze, for example, appears in blacks and a slight reddish tinge can be seen in the feathers particularly away from the tips. This is not a patch of blue followed by a patch of red but the gradual fading of one colour into another.

This same colour change is found in mosaics where blue and red fade into each other. One scientist has called this colouring opal after the colour changes of the precious stone. Another term used more by fancy, rather than racing, pigeon breeders is almond. In this the body colour is a light buff, splashed and flecked with red or black. The almond type of splashing is completely different from the type of splashing found on red cocks in almost every loft. As a preliminary to the chapter on breeding let me say here that a pigeon's appearance, its breeding ability and its racing ability are all controlled by genes, that is why we talk about the science of genetics. Nearly every gene in a baby pigeon is part of a pair, one from the mother and one from the father. One of these is the dominant gene and the other remains hidden or, in technical terms, recessive.

When talking about dilutes a few pages back, I described how to recognize them but not where they came from. Neither parent of a dilute will look out of the ordinary, but one or both of them will possess the dilute gene. When they are paired, the hidden gene surfaces and a dilute is bred. Black splashing works in almost exactly the same way but with an extra stage. The parents are a red cock and a blue hen. Any red cocks produced from this pair will have the blue gene hidden but in this case not completely hidden. The scientists call it incomplete dominance but they know no more than you or I. The hidden blue gene appears as the black flecks on the red feathers, usually on the wing flights or tail feathers. Surprisingly it increases with old age, year by year, and still more surprisingly in some birds it spreads faster than in others. One thing for certain is that the old red stock cock will be heavily splashed with black. It does not apply to all red cocks but, as many a beginner has found to his delight, it is a valuable clue when trying to sex his young birds.

The Head and Eye

The head of a pigeon is something about which surprisingly little is known, particularly as it is considered that inside the head lies the seat of the homing instinct. In view of some of these theories, the position of the eye in the head is most interesting. The eyes are placed either side and they look both forward and back so that a pigeon has almost complete all-round vision. It is only in a narrow sector right at the back of its head that it cannot see. It is only in a quite small area directly in front that both eyes can see the same object simultaneously. In the judgement of distances it is necessary for both eyes to be used, consequently it would seem that this has been sacrificed so that the bird can have all-round vision.

Experiments have been made to find out how good a pigeon's eyesight is and it has been found to be considerably better than that of a human being. Its particular excellence lies in the fact that a pigeon is able to pick out the movement of objects from very considerable distances. Some fanciers have used this ability to help them at the end of a race. The bird is trained so that as it flies over in a flock of pigeons, it will drop on to its loft the moment the fancier pulls out his handkerchief and flicks it. It is not known how a bird is able to detect such small movements, but since the eye of a pigeon differs in some respects from the human eye, it is thought that in some of these differences may lie the reason for its specially acute eyesight.

The eye is basically a globe with an opening, the pupil, on one side. Just behind this opening is a crystalline lens. This can alter its shape to enable objects to be focused on the back of the globe. The back of the globe is the retina. This is light sensitive, being covered by many thousands of nerve ends, each of which is able

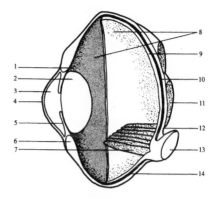

7 A PIGEON'S EYE.
1, ciliary body; 2, lens; 3, chamber of the aqueous humor; 4, cornea; 5, iris; 6, sclerotic plate; 7, pecten; 8, chamber of the vitreous humor; 9, retina or photoreceptive layer; 10, foveal region of retina – area of acute vision; 11, eyeball muscle; 12, sclerotic coat; 13, optic nerve; 14, chorioid layer or pigment layer.

to send a nerve response to the brain. The brain is able to correlate these nerve impulses into what we call vision. At the back of the eye is a small rod known as the pecten, not found in the human eye, and scientists have found it difficult to discover the exact purpose that it serves. It has been suggested that it is the shadow cast by this pecten that enables the pigeon to pick up the very small movements at considerable distances.

In front of the crystalline lens is the iris with the central hole of the pupil. The iris contains a circular muscle by which the pupil can be made larger or smaller, enabling the eye to admit less or more light as necessary. The iris of a bird is coloured and there are many possible colours. The commonest eye colour is orange, which at times verges on red or brown. More rarely it will be found to be yellow. An eye which is quite often to be found in white and gay pied pigeons is the bull eye. This looks like a completely black eye but if examined closely will be found to be very dark red. Another eye colour is the pearl eye which is quite frequently freckled, but is mostly a whitish colour.

Eyesign

Round the edge of the pupil (in other words on the inner circumference of the iris) is to be found a very fine line. This is the circle of eyesign, the circle of correlation or the circle of adaptation as it is variously called. One 'authority' has also called it the ciliary muscle but this is manifestly incorrect as the ciliary muscle is the muscle which alters the shape of the crystalline lens and is situated well inside the eyeball. Some authorities have drawn up a table showing the various shapes and colours that this circle may take and have graded these to show the racing ability and quality of a pigeon. From this foundation they have built up a whole system of pigeon management.

Although the theory has been elaborated considerably, those who put it forward cannot be said to have proved its usefulness. Several pamphlets have been written on the subject but not one of the authors of these pamphlets has had any outstanding success in open competition, and this must necessarily count against the theory they propose.

In recent years there has been a considerably increased interest in eyesign because the *Racing Pigeon Pictorial* has regularly printed colour photographs of pigeons' eyes. I am quite proud of my photographic ability in devising a practical system to take the pictures and I was the first to produce the picture of an eye enlarged approximately six times so that it filled a 35 mm transparency. I don't have time to take eye pictures any more but there are quite a number of people, including the famous Anthony Bolton, who do the job better than I did in the early days. The result of these photographs was that for the very first time it was possible to examine pictures in detail so there could be no argument about what was or was not to be seen in the eye. Quite often in the past, one expert would claim he could see one thing in an eye while another would assert exactly the opposite. Now, using photographs, it is possible to have an exact comparison, and the new generation of eye experts actually say that in many cases they find it easier to work from a photograph than from studying birds in a loft.

It seems significant to me that in spite of all that has been said and written in the *Pictorial*, the theory still remains a theory. My own objection to this theory is that, as in the case of wing theories, such ideas seem to approach the matter in the wrong way, starting at the end instead of the beginning. One of the reasons for introducing colour photographs of eyes was so that the eyes of champions could be seen by all. Any theories must develop out of these champion eyes and not vice versa. Nevertheless, the study of the eye does provide a great deal of pleasure to many fanciers and it still remains neither proven nor unproven that any of the eye theories actually work.

What worries me most of all is that no two experts can agree on a theory and that although some champions will pay lip-service to eyesign I cannot think of any great fancier who was dependent on eyesign alone. I realize that I may have lost some friends by saying this, and some of my best friends are eyesign men, but I feel it up to them to prove they are right, and not the other way round. Having got that off my chest, I have to admit that I look at the eyes of all pigeons when I handle them.

Homing Ability

It would need a whole book to go into a long and detailed discussion of the various theories on the homing instinct, so here I will just mention the possibilities. Homing instinct is a term used to describe the mysterious ability of a pigeon to find its way home over long distances. Some people suggest that there is no mystery at all and that a pigeon in its training learns to recognize certain landmarks by sight and to use them as signposts on its way home. If it cannot see a landmark then it circles round and round or wanders backwards and forwards until it does.

This theory will not bear any close examination. Untrained birds are quite often lifted for their first training toss to 15–20 miles (25–30 km), far beyond where they would have reached except by a lucky fluke when flying round home. The speed with which they return shows that they have wasted no time looking for the way, but knew it. Thousands of fanciers have bought old birds that have escaped from their loft before they have been settled and these pigeons have performed wonderful feats over unknown country. In America one bird flew 1,300 miles (2,090 km) on a flight such as this. Wartime pigeons regularly carried their messages hundreds of miles over unknown country.

Youngsters, even in their normal racing, at the later stages receive a lift of 100 miles (160 km) beyond their farthest point but the whole lot of them make straight for home. As a final example, I once watched the liberation of some Belgian birds in Battersea. Not one of these pigeons, their parents or grandparents had ever been across the Channel before and they had very little experience along this particular line of flight. Yet on being liberated they went straight as an arrow back to Belgium at high velocity not even bothering to circle once before they went off!

One of the theories that has always been popular has been that they are guided by the earth's magnetism and many experiments have been performed to demonstrate this influence. Birds have been let off near wireless stations that are transmitting at the time. Birds have been released to fly home with magnets attached to their wings, their heads and their tails. These experiments proved very little, for as fast as one scientist finds a change in the homing ability, another scientist shows conclusively there is no change! The magnetic theory is now generally discounted.

A third theory suggests that pigeons are sensitive to the rotation of the earth and can measure forces set up by this rotation. It is usually suggested that this measurement is done in the semicircular canals of the ear. Unfortunately for this theory, pigeons have shown that they can home quite reasonably even after all sorts of operations have been performed on the ear. From all the experimentation carried out by scientists there is very little agreement.

Some of the most valuable experiments have been done by scientists at Cambridge, particularly Dr G. V. T. Matthews. In the course of their experiments they examined all previous theories and by practical tests found them unsatisfactory. One of the reasons shows the gulf between practical animal keepers and scientists. Dr Matthews contacted a local fancier, Percy Cope, and he in turn spoke to my father and as a result every pigeon used in his tests came from stock that had been raced. You may well ask what birds had been used for previous experiments? The answer is that they came from specialized breeders of laboratory animals and probably bred from generations of laboratory animals. No wonder they did not give useful results for the earlier theories.

I have a theory that some pigeons bred from raced stock are bred without a homing instinct. I can't prove it but how else can you account for the fact that some young birds given exactly the same chance as others are lost off the roof? It is obvious that with all birds the sense of home is imprinted on their mind at about four to five weeks because if this were not so you could never buy a squeaker to bring in to the loft. Does it take longer for some to become imprinted? I don't know. I do know that well-bred youngsters can be lost off the roof. If I am right about some birds losing or never having a homing instinct, then laboratory birds could easily be bred from generations without any homing ability at all. No wonder the scientific experiments proved nothing and how right Dr Matthews was to go for racing stock.

This brings me to another topic – very expensive stock birds. Some of the biggest stud farms in this country have paid £30,000 or more for a pigeon. It is paired to one of the best but the progeny will be priced perhaps at £5,000 each and so will be kept for stock also for the same reasons. The same will apply to the next generation priced at £2,000 and the next at £1,000 and the next at £500. By the time the price has got down to a level where a fancier will put the bird on the road it could be ten generations of untested birds. If my theory about Dr Matthews's birds is right then among those ten generations could be birds without any homing instinct at all and so, whatever the money, they are useless for racing. What makes me fear that I am right is the apparently low success rate of the highly priced breeders. If only six rounds are taken from them each year then after ten generations there are nearly 9 billion youngsters descended from the first year mating; add to this the second and third year mating and my pocket calculator gives up. In schoolbook maths it is 12^{12} + 12^{11} + 12^{10} + 12^9 + 12^8 etc. The figures are astronomical but my question is very down to earth: where are the winners? I read the papers, I read the adverts but what have these £30,000 pigeons bred? If there are not even a million winners then the results are below average. Could this be because as the birds are untested on the road some of the negative genes controlling the homing ability have not been eliminated and have therefore been passed on?

The lesson is simple: stock birds bred from stock birds bred from stock birds may have built up too many negative factors to make them worth having. When it comes to the chapter on 'Breeding Winners' I will try and show how to breed the maximum of positive factors and the minimum of negative ones. We may seem to have wandered a long way from the homing instinct but we have not because all the parts of pigeon racing are related. You may for example choose a certain style of nest box in your new loft. I will be showing how this choice will probably affect your racing method, breeding method and even your feeding. This inter-relation always begins with the homing instinct, because without this there can be no pigeon racing.

Sun-Arc Navigation

Dr Matthews, in his book *Bird Navigation*, states his belief in a theory that pigeons navigate from the sun's position. Briefly, the theory states that a bird is able to visualize the path of the sun through the sky. The bird is able to judge where the sun will be at noon on this arc by means of an 'internal chronometer'. At noon the sun is, of course, due south in the Northern Temperate Zone and at its highest point. By measuring the angular height of the noon position the bird can tell whether it is north or south of its home loft. By measuring the difference of noon local time at the release point and noon local time at its home loft, the pigeon is able to tell whether it is east or west of its home loft.

Some fanciers have argued that although a pigeon is quite intelligent, no bird could possibly make a complicated mathematical calculation such as this would require. This is not necessarily so because all animals can make very precise measurements and calculations automatically. An example is the way it is possible to distinguish one shade of red from another shade. To do this the human eye and brain measures the wavelength of a beam of light far more quickly and more accurately than would be possible even on a most expensive micrometer. Nevertheless this theory is not without its snags since, although it may explain how a pigeon can find out where it is, it does not explain how it can find its way from this place to its home. The navigator of a ship or an aeroplane will first establish his position and after that he will lay off his course to his destination. The theory of using the sun does not explain how this course is laid off. The other snag that will have to be settled is that out of all the experiments conducted, none do more than indicate the possibility of the pigeon's internal chronometer. The existence of an ability to tell the time without using the sun has yet to be established beyond all doubt, but with continuing experimentation it may yet do so.

The ESP Hypothesis

There is yet one other major theory to be considered, although perhaps theory is too strong a word for it. This is the hypothesis which is based on ESP or extra-sensory perception. This takes us into the realms of fantasy because few facts are known for or against this theory, but it deserves some consideration. One is tempted to hope that it is right because of the ease and simplicity with which it works. Nobody is asked to accept the theory as proved and it is given here mostly for its interest.

Water divining by means of a bent hazel twig or a pendulum has long been known to exist. I myself have done it and most people given a hazel twig and shown how to hold it can dowse for water. Among the other curiosities of nature, some of which have been proved to all but the firm non-believers, are telepathic twins, telepathy of various sorts and even the detection of diseased tissue by the dowser's pendulum. Not everyone accepts these as proved facts and anyone who wishes to learn further on these should read some of the vast number of books on this subject. The theory behind these curiosities is that every living thing can be identified by its own unique wavelength. This wavelength is something similar to the wavelengths of light, infra-red radiation, X-rays, radio waves and even the radiations from atomic bomb explosions. The wavelengths of substances cannot be measured by ordinary instruments, but they can be detected by various devices. The hazel twig is the detector of the 'radiation' of water and with its aid ordinary people can discover water under the ground. Telepathic twins, being almost identical, have almost identical wavelengths and therefore quite often the thoughts of one quite naturally transfer into the mind of the other twin.

In the same way the pendulum can be used to detect water and can also be used successfully to detect the difference between live tissue and diseased tissue. I have met a woman who makes her living by detecting cancer in its early stages. There are many, of course, who would call her a quack and the whole of the theory a lot of tomfoolery, but this woman found on me an old concealed knee injury by her methods. The theory on which all of these are based goes on to explain that a pigeon loft has a compound wavelength which the pigeon can detect. This, in effect, is like a radar beam and the pigeon homes on this radar beam by flying down it to its source. It is as simple and as easy as flying directly towards a light.

Unfortunately, very little scientific work has been done on this theory, so therefore nothing can be demonstrated for or against it. Its important difference from the others is that it states that a pigeon homes to its loft and not to a geographical position on the surface of the earth. It does not therefore require the double calculation of finding out the bird's present position and then finding its way from that point home. The simplicity of the theory should not be allowed to obscure the fact that the proposers can produce no proof whatsoever to support it and the theory in itself is so vague that it is almost impossible to dispute it. Nevertheless, those who believe in parapsychology will find it of great interest.

TWO

The Pigeon's Year

*Cocks and Hens – Courtship – Mating – Nesting – Time of Laying –
The Breeding Cycle – The Importance of Pot Eggs – Pairing Up –
Fighting – True Pedigrees – Egg Laying – Infertility – Incubation –
Hatching and Pigeon's Milk – The Moult – Dropping the Flights –
Young Bird Racing and Growth*

One of the problems confronting a beginner is the difficulty of telling the sex of a pigeon. There are no visible sex organs nor are there any of the convenient differences in colour that are found in many other birds. Such differences as exist are relative and consequently it is quite possible for even the most experienced fancier to make a mistake. As a rule, of course, the fancier will be able to distinguish the different sexes in his own family of pigeons without much difficulty, but may find some difficulty with a type of pigeon he is not used to.

It is with young birds that there is more uncertainty, partly because they may not be fully developed physically and partly because they may not be showing any behavioural differences yet.

Cocks and Hens

In general the cock pigeon is larger than the hen, even in the nestbowl. Even this difference is by no means always a sure indication. There is also a difference in the head, but here again it is only possible to be vague about the difference. The cock has a domed head rising above the level of the eye whereas the hen has a shorter and more rounded dome and the eye appears to be set higher in the head. The cock has what is called 'the bold look' which seems more masculine than the hen, but as in all these cases, it is only a relative matter and no absolute rules can be laid down. These differences of the head and look are sometimes possible to see even in unfledged youngsters.

My old friend Taffy Bowen showed me a method that he had found infallible. I have tested it myself but whether it applies to all pigeons of all families I do not know. I would certainly like to hear from anyone who has tried it and found it not to work. It is very simple. You hold the pigeon in the usual way with the head towards you. Then with your free hand you fold down the rear toe of the foot so that it lies under the other toes. It goes easily because it is a natural movement. It is much easier to demonstrate than to write about. If you think of the backward facing claw as the thumb toe and the other claws as toes it becomes a little easier. Holding the bird so that the bottom of the foot, the pad, is visible, press the thumb toe gently back along the other toes. If the thumb goes naturally across the pad and between the outer toes the bird is a hen. If the thumb goes down the side of the pad and along one of the outer toes then the bird is a cock. Taffy Bowen, one of the oldest and most experienced fanciers in the sport, swears by the system and believes it infallible. Another way to decide the sex from the toes (and even easier because you do not have to handle the bird) is to look at the three toes (not the thumb). If the outer toes are equal, then the bird is a cock. If one of the outer toes is longer than the other outer one, then the bird is a hen.

Another guide which can be helpful is the colour of the plumage. Those who have a thorough grasp of the scientific principles and who know the history of their birds, can often work out the sex from the colour by following the scientific laws. This, of course, is possible only in certain cases. It is used in commercial pigeon breeding as auto-sexing is used by poultry breeders. Certain matings are made which produce all cocks of one colour and all hens of another. This, of course, is far more important to those breeding squabs for food, than to the racing fancier. An immediate guide which may be of help is that normally all mealies or red chequers that have black flecks in them are cocks. I always ring the largest youngster in the nest, hopefully the cock, with an even numbered ring and the hen with the odd number. I find it works fairly well.

The best guide and the one about which there can be no doubt is to find the bird that lays the eggs. If a hen is examined not long before or after laying, the increased separation of the vent bones can be felt by the fingers. It does sometimes happen that two cocks pair up together and sometimes two hens will. In these cases there will be either no eggs or four infertile eggs. In these unnatural pairings, one of the birds will usually act like its opposite sex during the courtship.

I should mention that the late Dr Stovin showed me how to examine the sex organs of a pigeon to discover whether it was male or female; there are even books which include photographs showing the difference. Very briefly: if you pull open the cloaca of a pigeon, in the male there will appear to be one small lump and in the female two small lumps. These organs are not particularly well developed in young birds and you need to be a scientist rather than a practical fancier to be able to do this without harming the birds. Some day we shall have a simple gadget which will enable us to examine them easily, but until then I will stick to traditional methods.

In *The Racing Pigeon* gadgets are offered quite cheaply that are claimed to sex pigeons accurately. I have not at the time of writing tried them so cannot report. I wrote about dowsing some years back and one form of dowsing does not use the traditional hazel twig but a pendulum. There are some who claim that this can be used to sex pigeons, but again I have not tried it.

Courtship

Just as in all other aspects of a pigeon's life, so in the courtship there are considerable variations with individuals. Some cocks are more ready to commence courtship than others and some hens seem to be almost frigid. Nevertheless, although there are variations in the intensity at which the courting actions are carried out, a general pattern is followed. If the birds are given a free choice, a cock will select a hen and his courting will begin with a proud masculine strut. He parades up and down in front of the hen 'blowing out his crop' and cooing with a deep rich coo.

Although the action is always called blowing out the crop, it is not the crop itself which is filled with air, but the air-sac which lies over the breast of the bird. While he is parading in front of the hen he will bob his head up and down and turn little dancing circles in front of the bird of his choice.

The hen, if unmated, does not usually resist his advances. If already mated, her cock will soon come to attack the interloper. The unmated hen, while not making any strong signs of acceptance, will indicate her readiness to the cock by a softer cooing and by a more restrained walk. She too will dance round the loft, walking slowly and leading the cock on. At this time it is usually quite easy to distinguish the sex of the birds and a quick guide is that in this courtship dance the cock coos with a deep note and in his dance turns a complete circle. The hen has a gentler, more feminine coo, and will rarely turn in full circles.

When this stage of the courtship has been completed the 'wedding' is sealed with the kiss of billing. This usually follows soon after the courtship dance. The cock will snuggle up to the hen and open his beak. The hen will put her beak into the cock's. During this kiss some people believe that the cock actually gives the hen some food by regurgitation, but whatever it is, it is only a token gift. The third stage of the courtship is the actual treading of the hen and although the action has been seen in two birds of the same sex mated together, this is unusual. When the hen is thoroughly roused and eager for a cock, she is said to be 'rank'. She will frequently follow the cock round as though pleading for a kiss. When the cock stands still she will run her beak through the coverlets on the back of the cock's head and neck as though stroking him, at the same time making a nibbling movement with her beak. Soon after this, copulation usually occurs.

Mating

At the climax of their love making the hen will offer herself to the cock. She crouches down and separates her wings very slightly. The cock will mount her by treading on her back and as the hen turns slightly to one side, the sexual act is completed. This act takes only a fraction of a second, when the two sexual openings touch briefly. In the case of older birds which are more experienced in the courtship there may be a reduction in time taken over the preliminaries but this again is a matter which varies considerably in individual birds.

Distinguishing sex from the courtship is a comparatively easy matter during the breeding season, but even when the birds are separated it is possible to watch them courting the birds of the next compartment of the loft. The cock will coo and turn a full circle on his perch whereas the hen will make a smaller movement and coo back more gently.

After the mating, the birds will begin to go to nest. If the pair has not already been given a nestbox, they will find a place for the hen to lay her eggs. The cock usually finds a suitable spot and then begins driving the hen towards it. After the tenderness of the courtship this can seem surprisingly cruel. The cock will chase the hen round and round the loft pecking at her head, and even buffeting her with his wings, until she goes into the nestbox. This is called driving and its apparent viciousness may have a real purpose. The hen is being driven to the nest so that she will be out of the way and unavailable to other cocks. A rank hen, feeling sexy, would accept the attentions of another cock. By driving her to nest, the cock is making sure that he will be the parent of the eggs that are laid. It could be a throw-back to primitive times when this was a way of ensuring the survival of the family, and to support this idea it is less likely to happen so violently with two stock birds on their own or in an uncrowded loft.

Nesting

Wild pigeons do not bother much about their nest, which is rarely more than a few bits of straw or feathers gathered together. Inside the loft a fancier will normally provide a nestbowl for the birds. This is usually made of unglazed earthenware. The advantage of this type of pottery is that it is porous and therefore will tend to absorb a certain amount of moisture.

This is beneficial to the hatching of the eggs. At the same time the bowl is not too cold to the touch for the bird.

Other substitutes have been suggested; one fancier swears by turned wooden bowls which, although very high in initial cost, are virtually unbreakable and are popular with his birds. I did have one or two but did not get on with them because they seemed to hold the dampness from the babies' droppings and with that dampness any disease. Most popular because of their cheapness are the plastic bowls. Some I think are too smooth inside; I like them ridged so that any nesting material does not slide about but stays in the bowl. Some fanciers swear by nest-bowls made of papier mâché; Dandynets are the best known. These have the advantage that they are thrown away after use, preventing the spreading of disease and mites through the dirt that accumulates on them. These nestbowls were, of course, quite cheap. Chaff or hay used to be put inside the bowls, but nowadays sawdust is used almost universally. This should be dusted with a mild insecticidal powder before the bowl is put in the nestbox.

Although a nestbowl is provided with sawdust inside, most birds will follow the natural habit of building and will enjoy gathering up bits of straw and twig to line their nests with. A handful of straw put in the loft will help to make contented pigeons. At the big shows like The British National (Old Comrades) Show some stalls have on offer tobacco stems. These usually cannot be obtained mail order because of their bulk but they make very good nesting material. A visit to the show is also the ideal time to stock up on my favourites, the earthenware nestbowls. These are too heavy and too breakable to be delivered so are ideal for collection at the show particularly as they are usually on the same stand as the tobacco stems.

Pigeons vary in the amount of building they do. Many pigeons will keep building nests all through the year and will industriously search out twigs and even feathers right up to the end of the season. It never seems to make much difference to a bird's performance whether it is a good nest builder or not.

It must be remembered that the second round of eggs will be

laid while the first round of youngsters are still in the nestbowls. The parents will solve the dilemma of where to lay the second round of eggs in one way or another and either lay the eggs in the same nestbowl as the youngsters or in a corner of the nestbox. The fancier can, of course, solve their dilemma by putting a second nestbowl into every nestbox when the cock begins driving again. If this is done the hen will almost invariably lay the second round in the other nestbowl.

Perhaps this is a good place to stop and mention the cleaning out of the nestboxes. The baby pigeons have an instinct for cleanliness and it always surprises non-fancier visitors to see the way youngsters often only a week old will push themselves backwards up the inside of the bowl and defecate over the edge. The droppings are very soft and in hot weather can become very unpleasant, smelly and soon crawl with maggots which become bluebottles. It happened in my loft once during a time when I was ill. Never again! The nestbox round the bowl must be cleaned daily. I have recently bought some plastic corner shields and I have every hope they will help when cleaning the corner. This is why I like pottery bowls; it is easy to lift them up, go under them with a little shovel and put them back. With papier mâché this is not quite as easy or quite as fast and the whole point is that it must be done in a way that disturbs the young babies as little as possible.

Birds that build big nests can be a bit of a problem. I had one that would never use a nestbowl but found a corner of the loft and started to build a nest with twigs, adding layer by layer until the nest was a little tower nearly 10 in (25 cm) high. I know another fancier with the same problem who removes an inch or so of twigs every day. I lost my bird racing which was just as well because the eggs never hatched. Either another bird would raid the nest and steal the twigs or the eggs would fall off the top of the tower. I only ever hatched and reared one round and that was when I surrounded the tower with bricks. Even then, because the nest was on the floor, I was not happy, because babies on the floor are always in danger of being scalped.

For the novice loft-owner scalping is an unexpected risk. A

young pigeon, often unable to fly up to the nestbox it has fallen out of, finds itself running around the floor and attacked by adult pigeons, which will peck at its head and quite often tear a piece of skin and feathers. The attacks are well named scalping but why this attack takes place is still a mystery, but needless to say I have a theory that, like driving to nest, it goes back to primitive times when a colony of pigeons would be competing for limited amounts of food. A baby pigeon was therefore a threat to the food supply and to be attacked when out of the protection of the parents' nestbox. Whatever the explanation, if a pair does nest on the floor the only way to protect any babies is to give the parents some 'territory'.

The most obvious example of territory is the nestbox. Once the cock and hen have settled in, that box is their territory and woe betide any trespasser. Some cocks may even lay claim to an empty nestbox to extend their territory even though their hen is in the first box. This must not be allowed because sooner or later it will lead to a territory challenge and fighting. There must never be more nestboxes open than there are birds to fill them. Some birds go further and lay claim to part of the floor. Annoyingly this is sometimes the food hopper or the drinker and one aggressive cock will try to prevent any other bird feeding or drinking. This usually does not work because the hungry birds will gang up on him and make his task impossible.

When a pair nests on the floor you have to create a territory. This can be done with bricks. A semicircular wall about 12 in (30 cm) away from the nest forms an easily defendable territory – even a very aggressive cock will not venture into it. Do not underestimate the importance of territory because, as we shall see, it is very important with the Widowhood system, for many successful Widowhood fanciers will tell you they have won races with the cocks racing to the nestboxes and not seeing the hens before and sometimes not after the race.

Nowadays the most popular method of racing pigeons is the Widowhood system. Most of my readers will be waiting to get on with the chapter on winning with Widowhood. I do urge patience. Pigeon races are won by patience and in this case, before you can win with Widowhood, you have to understand the alternative Natural system. It may be a bore but you don't have to race the birds Naturally. The reason for this is that although you can argue that the Natural system is not really natural, it works because it takes into account the normal lifestyle of a pigeon. The Widowhood system by comparison, because it interrupts the natural cycle, is regarded as unnatural. Before we get carried away by the words natural and unnatural let me point out that in the wild state pigeons do not race, they do compete with each other for food and nesting sites but you can not imagine one pigeon saying to the others 'Come on, lads! Let's go for a race!' Mind you, the same thinking makes those humans who go jogging just as unnatural. We do not live in the jungle any more and adapting lifestyles is part of the compensation we make. If we swung from the trees, we would not need to go jogging. In the same way, just as our lives have been altered so have the lifestyles of pigeons. They no longer live in caves but were domesticated centuries before pigeon racing started. You will notice that to make the difference more obvious, the Natural system has a capital 'N', but when describing something I think is really natural I use lower case. Let us make a start by looking at the natural life-cycle of a pigeon.

Time of Laying

The laying of eggs is timed, for the first egg is normally laid between 4 and 5 p.m. Variations may occur, but it is rare for the first egg to be laid in the morning. My loft has electric light and if this is used to extend daylight, then the first egg is laid later. The second egg of the pair is laid the next but one day and this egg is nearly always laid nearer midday. The time interval between the first and second is about 44 hours. It can be seen that if a bird is taken to the race marking and lays on the way in the basket, since this is the afternoon, the second egg can be expected on the morning of the next day but one. If the marking is on Thursday this will be on the Saturday morning and even if the bird were in the best physical condition to fly it would pitch so as to find a suitable place to lay its egg. It may even sit this egg for a few days. For this reason if hens on the point of laying are sent to a race they are very rarely winners and quite frequently lost.

The Breeding Cycle

The whole of the breeding cycle of pigeons follows a fairly regular timetable. About 5 days elapse from the time of mating to the time when the cock begins to drive the hen hard. The hen will lay about 5 days after the driving has begun, that is 10 days after mating. The second egg will be laid 2 days, or more exactly 44 hours, after the first egg. These eggs will take 18 days to hatch, that is 28–30 days or 4 weeks from the time of mating. About 10–14 days after this, the hen will lay the first of the second round of eggs, that is 38–44 days from the time of their being mated. The customary warning must be added that although these times may be true for most pigeons, every pigeon is an individual and its breeding cycle can easily have variations. Quite often 3 weeks will elapse before the second round is laid.

The table overleaf is a Natural racing table which is why it begins in March. Early breeders will normally pair up early in December and January and then separate the birds. The table shows when they should be allowed to repair for racing. The system can be speeded up by removing the eggs earlier or slowed down by using pot eggs and waiting until they are deserted. As we shall see shortly when talking about the moulting of feathers, the moult does not begin with either the cock or hen until the second egg of the second round is laid. It does not matter whether this second round is in January or April as it will still be roughly true, but as spring turns to summer it becomes less and less true. We shall see later how Widowhood and long-distance fanciers use this fact to their advantage. To delay the moult the birds must be separated again; the use of pot eggs controls only hatching and feeding.

The Importance of Pot Eggs

When the second round has been laid it is not uncommon for those probably fertile eggs to be removed and pot eggs of china or plastic to be slipped into the nest. By doing this a racing pair will be saved the strain of continual feeding of youngsters and their strength will be conserved for racing. If they are a valuable breeding pair, the eggs can be incubated by another pair of lower racing value. The hen of this pair should have laid about the same time and these foster parents can be used to rear what is hoped will be valuable youngsters. Both the cock and hen of the racing pair will continue to sit on these pot eggs until the time when they should hatch and soon after this the cock, followed shortly by the hen, will desert them. After a few days they will begin to drive again when the same procedure can be repeated.

China eggs is only an expression; years ago they really were pottery (i.e. pot) eggs and I liked them because the weight of the egg was about right. Plastic eggs not only seem too light to me, but the ones I use discolour easily. I am more and more convinced that the cheapest solution may be the best and that is to get some freshly laid eggs and hard boil them. They must be cooked because I have known some fanciers use old eggs months after they have been laid and found to their surprise that they have come to life again and produced youngsters.

I once did some experiments with eggs to find the best way of sending them overseas. The conclusions I and the far distant fanciers came to were that the eggs should be left about 2–4 days and then taken away and allowed to cool. They could be left a few days longer but 6–8 days was the maximum. For transport we first tried sawdust, but this was a failure as the eggs worked their way through the dust to the edge of the box where they were vulnerable. In the end we settled on sheets of cotton wool although I suspect that newspaper would have worked. I have been told that, like chicken eggs, they should be turned regularly and we started by urging aeroplane travellers to turn them every day by simply turning the box upside down. We then found that eggs sent through the post (where legally permitted) come to no

The Natural System

The table cannot be guaranteed for all birds as there will be considerable variation depending on conditions, but is intended to give a rough guide for the less experienced fancier.

How to find from date of mating

Date of Laying	Add 11
Date of hatching	Add 29
Date of laying again	Add 43

These figures should be added to the table number of the date of mating.

Example

Birds mated 6 April	Table number 37
Date of laying again	Add 43
Total	80 (19 May)

How to find date of mating so that the bird is in the right position for racing

Best Position	*First Round*	*Second Round*
Driving	5	38
Point of Lay	10	43
Sitting 3 days	13	46
Sitting 7 days	17	50
Sitting 10 days	20	53
Sitting 14 days	24	57
Hatching	28	61
Young in nest 3 days	31	64
Young in nest 7 days	35	68
Young in nest 10 days	38	71
Young in nest 14 days	42	75
Young in nest 21 days	49	82

These numbers should be *subtracted* from the table number of the race date – result gives table number of date of mating.

Example

Date of race, 8 July	Table number 130
Bird's best position is three-day old youngsters	Subtract 64

Result: 130 minus 64 equals 66 – 5 May
Birds should be mated for previous round 5 May

March	Table No	April	Table No	May	Table No	June	Table No	July	Table No
1	1	1	32	1	62	1	93	1	123
2	2	2	33	2	63	2	94	2	124
3	3	3	34	3	64	3	95	3	125
4	4	4	35	4	65	4	96	4	126
5	5	5	36	5	66	5	97	5	127
6	6	6	37	6	67	6	98	6	128
7	7	7	38	7	68	7	99	7	129
8	8	8	39	8	69	8	100	8	130
9	9	9	40	9	70	9	101	9	131
10	10	10	41	10	71	10	102	10	132
11	11	11	42	11	72	11	103	11	133
12	12	12	43	12	73	12	104	12	134
13	13	13	44	13	74	13	105	13	135
14	14	14	45	14	75	14	106	14	136
15	15	15	46	15	76	15	107	15	137
16	16	16	47	16	77	16	108	16	138
17	17	17	48	17	78	17	109	17	139
18	18	18	49	18	79	18	110	18	140
19	19	19	50	19	80	19	111	19	141
20	20	20	51	20	81	20	112	20	142
21	21	21	52	21	82	21	113	21	143
22	22	22	53	22	83	22	114	22	144
23	23	23	54	23	84	23	115	23	145
24	24	24	55	24	85	24	116	24	146
25	25	25	56	25	86	25	117	25	147
26	26	26	57	26	87	26	118	26	148
27	27	27	58	27	88	27	119	27	149
28	28	28	59	28	89	28	120	28	150
29	29	29	60	29	90	29	121	29	151
30	30	30	61	30	91	30	122	30	152
31	31			31	92			31	153

harm with the normal movement. Experimenting at home, I turned eggs weekly for 2 months and then put them back under birds on the nest and got a good if not perfect hatch. Others have told of keeping eggs 3 or even 4 months successfully. This all goes to show why chilled eggs don't make good pot eggs.

It can be seen that it is quite possible to prevent a bird feeding any youngsters at all during the year and in this way it might be thought that maximum strength would be conserved. This is only true up to a point since too great an interference with the natural breeding cycle will undoubtedly subject the bird to undue mental and emotional strain and it may be found to race badly and to be generally uninterested. It is, of course, possible to replace only one of the eggs of a pair so that only one bird is hatched out. This single bird will then receive the double attention of both parents and such single reared youngsters are highly esteemed by some fanciers, although they are by no means universally successful.

Another advantage in using pot eggs is that it is possible to regulate the time of laying to a limited degree. If the birds are sitting on pot eggs and these are removed after say 15 days then the next breeding cycle will begin that much earlier. In this way the breeding cycle can be adjusted so that the birds are in the right position on the nest for a particular race. The importance of this will be realized later after reading about the Natural system for winning races.

The final advantage of rearing one round and then using pot eggs is the gain it gives to the young birds themselves. To show this let us compare two lofts. In the first loft the birds are all paired up at the same time so that they will lay at about the same time and hatch out at the same time. The old birds are put on pot eggs and no more are hatched for 3 months. During these months, the youngsters are weaned at the same time, start to fly at the same time and can be trained at the same time. In the second loft, by contrast, the birds mate up as they please, lay and hatch at different times and continue hatching until at the end of 3 months there are youngsters of every possible age in the loft. As these are weaned they are separated into the young bird loft and there no sooner are they old enough to fly than they are carried off and lost by the older ones. This is particularly the case where the early-bred youngsters are flying off, perhaps going off for 3–4 hours at a time. Unless the fancier has more than one section for young birds, continued breeding will only give him huge numbers of young birds and many will be lost off the roof. I have not done it but Bob Hutton, who flies a good pigeon, lets none of his young birds out until the second round joins the first. In theory he should lose a lot of pigeons but in fact he loses very few. If it works for him it should work for others. The proper use of pot eggs is one of the most important things the novice has to master for successful Natural race management.

The most normal arrangement preferred by fanciers for their long-distance candidates other than the Widowhood ones is that they shall rear one or two youngsters of the first nest. Thereafter the birds sit on pot eggs for the remainder of the season and perhaps rear a pair of late-bred youngsters in the late summer or autumn. Late-breds are, however, a special problem which will receive consideration later. The racing season seems to start earlier every year and there are many people, like myself, who are not at all happy with this because it means that it is more and more difficult to rear from pigeons prior to racing. In fact, a number of prominent fanciers now compete in early races before they start to breed and then take a first round of only one youngster in an attempt to cut down the strain on the birds. It is still true that you cannot race and breed from the same bird and we are causing ourselves trouble if we try to do this.

The 1970s saw quite a spectacular growth in marathon racing in the United Kingdom; by this I mean races of 600, 700 and 800 miles (960, 1,130 and 1,290 km) which it was known could not be completed in a day. I must confess that my knowledge of this is limited even though I was privileged to be the President of the British International Championship Club, which arranges British participation in the races open to the whole of Europe from these marathon distances. The reason I don't compete is simply that they are South Road races and I fly on the North only.

The nearest we have to a marathon on the North Road is the

Lerwick race which can sometimes be flown in a day if there is a very early liberation in the almost Arctic north. In the North Road Championship Club there are quite a large number of successful fanciers who do not believe that you can rear and race from the same birds, and even some like the late Olive Bridge who carried it to the length that her Lerwick candidates were not allowed to rear any birds at all before being sent to the race. Continuing the theory still further, after they had gained experience down the road they were not raced extensively the year they went to Lerwick. On their return from the race, in order that the Natural cycle may be resumed, they were allowed to rear a nest of youngsters, and as the race results show through the years, this system worked most successfully for her.

This goes back to 1954 and was particularly memorable for me because her pigeon was the first successful eye picture I took. In November 1994 I was visiting Demeyere in Belgium and he showed me his Barcelona winner. This bird had not only raced and won at 800 miles (1,290 km) on Widowhood, but when shown to me still had three and a half flights to moult. The owner reckoned that it would not be finished until January and that based on previous experience the moult would be perfect. His method was the same – the cock was not paired up before racing and his only rearing was a pair of beautiful-looking late breds reared after he returned from the race.

Pairing Up

As a rule a fancier will not give his birds a free choice in the selection of their mates. He will want to select certain pairs on the basis of their performances and pedigrees with a view to breeding even better birds. He has therefore to persuade certain birds to pair up together. Sometimes there is no difficulty about this at all and the two, if put in a nestbox, will soon settle down to a life of conjugal bliss. Other birds are more intractable and unless great care is used will prove difficult to settle. The easiest matings are those where birds are remated in the same pairs as they were the previous year. Some of the most difficult matings are those in which a pair which has been together for several years is broken up and both birds are remated.

In mating up it is essential that fighting should be avoided and that the mating of the loft should be got through with the minimum of disturbance to the birds. The fancier must make up his mind to take his time about it and see to each pair of birds individually, only proceeding to the next pair when he feels certain that the first pair is going to cause no trouble.

The style of nestbox with what are usually called Widowhood fronts is among the cheapest and simplest methods. In the past these divisions were made of wood and were difficult to adapt. The latest ones are plastic and can be adjusted within limits. The plastic fronts have a plastic separator halfway across the nestbox so that with the cock on one side and the hen on the other, the birds soon get to know each other. Needless to say there must be no loose birds in the section of the loft but all the nestboxes can be filled, that is provided you have a drinking pot for each one.

The Widowhood front with central partition or even a removable partition that can be put in the nestbox is useful. The cock is placed one side of the partition and the hen the other so that they can see each other but are separated. They are kept like this for 24 hours or longer and fed on a mixture of heating seeds until the cock will accept the nearness of the hen. Then the partition is removed so that the birds can reach each other. This will usually be successful but if it is not then it is usually best to consider whether another mating is less trouble for the fancier. Fortunately these cases are rare.

All the birds in the loft are shut up either in their nestboxes or in a separate section of the loft. One nestbox only is opened and one pair of birds allowed into that section. If given time and quietness the cock will drive the hen into the nestbox and they will settle down there. It is sometimes necessary to shut up the cock and hen together if they do not settle quickly. In difficult cases if the birds are fed small seeds they may be encouraged to settle in the nestbox. The fancier must then wait and must watch carefully to see how they are settling down. Not until he is certain that there is going to be no fighting should he stop watching them.

Fighting

It cannot be emphasized too much that pairing up cannot be hurried and several days should be set aside for it. If fighting breaks out, the hen must be removed at once otherwise birds can be injured, and also the rest of the pairs will be disturbed. If fighting does break out with one pair, the hen must be removed and taken back to her own loft and her pairing left until last. On the next day, the fancier can try again with this hen, once more removing her if the cock is hostile.

One of the great dangers in a loft is having uneven numbers of each sex. An odd cock can cause considerable trouble since he will always be trying to chase a hen from a mated pair. If the number of birds of each sex is unequal then either the surplus should be got rid of or birds of the opposite sex obtained to make the numbers even.

True Pedigrees

If a fancier is mating together his two specially selected birds, in certain cases he will want to be absolutely certain that the sire of the youngster is the cock of the pair. Although pigeons are by tradition faithful unto death, in fact this is not so. The cock of a pair will usually prevent any other cock approaching his hen when he is there. But in his absence, for example at a race, the hen may allow another cock to tread her, consequently invalidating the fancier's pedigree without his knowledge. The only absolutely certain way of making sure of the genuineness of the mating is by having the pair separated from all other birds at the critical time.

Egg Laying

The hen pigeon begins egg production as a result of the glandular stimulation provided by courtship of the cock. The egg yolk is first formed in the ovary and for the next $4\frac{1}{2}$ days it continues to grow still secured to the wall of the ovary. A speck on the side of this yolk, the blastoderm, is the beginning of the embryo. On the evening or night of the fourth day the yolk, now full-sized, leaves the ovary and begins its way down the oviduct. It is during its passage down this oviduct that the yolk receives the albumen coating, the white of the egg, which forms the rest of the food until hatching. At the lower end of the oviduct the yolk and the white receive the glandular secretions that harden to form the shell.

The time taken for the passage of the egg down the oviduct is about 40 hours and fertilization should take place in the early part of this period, about 15 hours after the egg leaves the ovary, that is not later than 24 hours before the egg is laid. If the egg is to be fertile then a male sex cell (sperm) must make contact with the female egg cell. If the hen has been trodden in the 10 days prior to her fertile period then fertilization can still take place. For this reason a hen should be separated for a fortnight if a fancier wishes to be certain that there are no residual male sperms which could cause a hen to be fertilized by the wrong cock.

There is a difference of opinion among the scientists about what happens if a hen is trodden by two cocks and has therefore two different sorts of sperm capable of fertilizing her. Some maintain that the more recent and therefore more vigorous sperm will be those that fertilize the egg cell but not all scientists are in agreement with this. The only way to be certain is to see that the hen is separated from all cocks for 10–14 days. The passage down the oviduct is the same whether the egg cell has been fertilized or not.

Infertility

The customary 'cure' for infertility is to remate the birds into fresh pairs, since it may be due to the pairing up of either two cocks or two hens! Another cause for infertility can be the failure of the male and female sex organs to make the proper contact because of the amount of fluffy feathering on either one of the pigeons. This can be cured by trimming or plucking the feathers. There is a strong belief that vitamin E will help increase fertility. It is available as drops or tablets, which can be administered individually to a bird. I don't think that any research has been done on dosage but, as in the human world, mega-dosing seems to do no harm so it seems not to be critical. It is now more fashionable to give birds vitamin supplements regularly. Most of these contain vitamin E but I doubt if this would be enough to cure apparent infertility so a special vitamin E supplement is needed.

At the back of my mind is the nagging quibble that none of this treatment for infertility should be necessary. It is not likely to occur in birds until they are ten or more years old. Their best racing performance is most likely to be up to 6 years old which means that any fancier should have four years of breeding. That should be enough. If there is something fundamentally wrong with the pigeon, should one persist? The professional breeders will go on of course because it is money, but for the average sportsman is it worth it? Hereditary infertility is not likely to worry the fancier since any bird suffering from this weakness would probably be suffering from other weaknesses which would have been shown during the racing season.

Incubation

The hen does not begin to incubate the eggs until the second one has been laid. In this way the bird is able to ensure that both eggs hatch out about the same time. Inside the egg during the 18 days of incubation, the blastoderm changes into an embryo living at first on the yolk of the egg and later on the white of the egg as well. In addition the embryo needs air, which is absorbed through the wall of the shell. As long as incubation is continued, the eggs, provided they are fertile, will be hatched in the usual way. If in the first few days incubation is interrupted and the eggs are allowed to get cold this will not prevent them from hatching and when the birds begin to sit again the egg will continue its change into the embryo. If chilling should take place while the embryo is in its later stages of development, the egg will not hatch because the embryo will have died in the shell.

It is possible to check on the development of the embryo by 'candling'. If an egg several days old is held up to a bright light (nowadays a candle is rarely used since there are more convenient light sources) the embryo will be seen as a dark shadow. If the egg is clear, fertilization has not taken place. It is obvious that when birds are sitting, the eggs and the nestbowl should be handled as little as possible so that the birds are not unduly disturbed. Unlike the wild birds, pigeons will not usually desert their eggs if they are disturbed, but unless there is a very good reason, they should not be interfered with.

This does not mean the birds should be left entirely alone. Tameness in pigeons is such a desirable quality that no chance should be missed to promote it. Talking to the birds, stroking them and gently lifting the birds off the eggs to candle the eggs are all most desirable but the moment a pigeon shows signs of alarm, that is the time to stop. If the fancier puts his hand in the nestbox while a bird is sitting eggs that bird will buffet his hand with its wing and will often peck at the hand. This is normal behaviour, not what I would call signs of alarm. Alarm is when the bird leaves the nest instead of sitting tight. It is natural for a bird to defend its territory whether from another bird or a human. It is not natural for it to leave its eggs even temporarily.

The incubation of the eggs is shared by the cock and the hen, but the hen takes the greater share. Her turn of duty is roughly from four o'clock in the afternoon until nine o'clock in the morning, nearly 18 hours including the night, while the cock sits the remaining 6 hours. These hours are not rigid for every pair but obviously this is when, perhaps inconveniently, the hens should be exercised. During this period the bird that is sitting will not normally leave its nest for feeding unless it is extremely hungry. For that reason food should be available at least twice a day, once during each bird's turn of duty. This turn of duty is another way of telling the sex of a pigeon, but usually the fancier will have found this out before the birds go to nest.

Hatching and Pigeon's Milk

The actual hatching of the egg is performed by the embryo inside. On the tip of the upper beak a little horn-like knob grows.

This egg-tooth is used by the embryo to release itself from the eggshell and is then absorbed into the bird's beak. With this egg-tooth, the little chick pecks a small hole near the wider end of the egg, it then turns slightly round in the shell and pecks another hole. It continues this pecking until it is able to push the end off and emerge. Unlike other poultry, chickens in particular, the pigeon when it emerges is completely helpless and only just able to move. The parent pigeons are almost unique since they feed the infants on a secretion of the crop known as pigeon's milk. This substance is formed during the time the eggs are hatching, in slight amounts early on and then later in considerable quantities. It forms a thick layer round the lining of the crop and by the time of hatching weighs nearly 1 oz (28 g). The milk is formed largely from the actual walls of the crop which change considerably at this time.

Pigeon's milk is something which is unique to pigeons and doves among the more common animals. It is not, as in most other birds, a pap of regurgitated food. Since it is a product formed by the bird itself, it is in this respect similar to mammals' milk, although unlike mammals both the male and the female produce this liquid. The real justification for calling it milk is that it is stimulated by the hormone prolactin just as mammals' milk

8 LARGER FEATHERS OF THE WING.

is. The squeaker is fed on this milk alone for only about 3 or 4 days and then regurgitated corn is added in small amounts. The soft food is taken by the youngsters which reach right into the old bird's mouth to do it. This is the time when I think feeding pellets helps both parents and chicks. At the age of 2 or 3 weeks, the youngsters will be feeding almost entirely on regurgitated food and will get out of the nestbowls to chase their parents for food. They will also be beginning to pick up corn.

The Moult

The moult is the natural process by which the pigeon changes the whole of its plumage once a year. The moult begins with the innermost of the primaries, and one by one each primary is replaced to the end of the wing. This process can be seen by watching a bird and fanning its wing regularly. A feather will loosen in its socket and as the bird rises from its perch or nest-bowl, will fall to the floor. Within a few hours a sheath containing the new feather will begin to emerge from the socket. This cannot be actually seen without considerable difficulty as the quill ends of the feathers are concealed by the coverlets. The sheath is pushed outwards and very shortly the new feather begins to work its way from the sheath and outwards in the wing. This will work farther and farther outwards until by the end of a week it has taken the position of the old feather.

It is obvious that if a pigeon is to race at its best, it should not be asked to do so if a number of the outer wing flights are missing, particularly as it may well happen that one outer flight may fall before the one next to it is fully grown, thereby creating a big gap in the wing. For the long races ideally the moult should not have progressed beyond the third or fourth flight leaving all the outer ones unaffected. The moult is influenced by the breeding of the bird for the first feather will not normally fall until the bird has laid its second round of eggs. For this reason the fancier should mate the birds so that the laying of the eggs and the beginning of the moult take place at the right time of year.

Dropping the Flights

The moult in the primaries of the wing of the bird is very regular, and the flights almost invariably fall in a regular order. There are nearly 6 months between the fall of the first flight and the fall of the last. During the same period all the secondaries will also be moulted and so will all the tail feathers. Of the tail feathers the first to fall are usually not the central pair, but the pair that lies on either side of these. Apart from starting in this order there is no fixed sequence of moulting for either the tail flights or the secondaries, although as in the primaries, one flight will normally fall within a matter of hours of its opposite number in the other wing.

The interval between moulting flights can vary considerably, and after a severe race, or after any injury or other similar tax on the bird's strength, there may be some considerable delay until the next flight is moulted. The new feather which comes up will normally show the signs of this increased strain by a fret mark, a break in the webbing of the feathers where the barbs have not been able to grow normally.

Fret marks are often called the pigeon's medals since they indicate a more than usually hard race and a difficult flight. While a few will not have a great effect on the wing of the bird, a badly fret-marked wing will not increase the efficiency of the bird's flying and in the show pen few judges will take kindly to birds with heavily fret-marked wings, even in the flown classes.

From what has been written it is obvious that the moult and the breeding cycle of a pigeon must be considered most carefully before sending a bird to any race except the shortest and easiest. One of the secrets of a successful Natural pigeon racer is that he makes up his mind which pigeons he is going to send to certain races and prepares them for it. He maps out a plan of campaign so that they will be just right when the time comes. He must have them at the right position on the nest so that they are at their keenest to race home and he must have them in a good state of feather. The questions involved by the position in the breeding cycle will be considered later. The state of the moult is something which must be planned right from the beginning of the racing season.

The usual idea is to delay the moult as much as possible by natural means. The most obvious of these is by mating the birds up fairly late in the season. There are other things which will speed the moult and they include not enough exercise, too rich a food and too warm a loft. Providing these things are avoided then the bird should have a satisfactory wing for the races in July. Families of pigeons vary in the speed in which they moult. Some drop their flights rapidly and some slowly. In addition to this there may be delay due to particularly heavy demands on the bird's system. The two most important of these are hard races and feeding youngsters, both of which will delay the moult.

Although most fanciers pay great attention to the moult of old birds, particularly on the long races, with youngsters very little attention is paid to the number of flights carried. I have seen youngsters sent 200 miles (320 km) with four tail feathers missing. The regrowth of these flights must place a very considerable strain on the bird's system and if that bird has a hard race the strain on its internal system could so check its growth that it might never make up lost ground.

Young Bird Racing and Growth

The question of whether to send birds to the limit as youngsters is one on which there is very considerable discussion. Most fanciers will agree that young bird performances are no guide to what the bird will do as an old one. As an illustration, very few of the winners of the Young Bird National ever turn out to be highly successful as old birds. Although most fanciers are agreed on this, many argue that a pigeon, especially a hen, must fly the whole distance in the season otherwise there is no means of telling whether or not the bird is worth keeping.

As opposed to this, there is another school of thought that says that young birds should not be raced at all but allowed to develop without being worked. The real solution seems to lie somewhere in between. The moult in youngsters is more rapid than in the same birds when older and, once the birds begin to moult heavily, there is an immense strain on their system. It is then that they should be stopped otherwise the foods which should be producing extra muscle and new feathers will have to be used in mending the old. However, if the birds are not raced at all then the muscles will not be exercised sufficiently to enable full development.

My plan to follow then is to race them at all stages until the moult is well advanced. They should not be raced bald-headed or with half a tail. The body moult can occur early in the young bird season; if this is so it is best to miss a race or two and then jump them to the next race point rather than send them loose in the feather. This, of course, only applies to well-trained youngsters. Even more than the old birds, youngsters should be allowed to rest after a hard race, even if it means keeping them in for a day or two. A good moult is a sign of a good bird, and sound feathering is a sign of good muscles. Of the greatest importance during the moult is the food on which the birds are fed. It must be nourishing and it must be the right sort of food. This will be discussed in the next chapter.

The Belgian professionals, and some of the British, go to great lengths to get the moult right. If you are only going to race hens as young birds then the YB season becomes very important. Normally bred youngsters are becoming very ragged by the end of the season. The flights are falling and the body moult that starts behind the nose wattle has spread to the head such that they are often said to be racing bald-headed. To combat this YB specialists breed as early as possible, and have birds ready to ring on the day that the official bodies release the rings. In England it is 10 January, in Belgium 1 January, and in Wales December. This means the birds must be paired in December or even November. The benefits are that these January-reared youngsters have moulted out their juvenile plumage by the time of the races and are not flying bald-headed but with a full wing, which gives them an obvious advantage.

The disadvantages are that if these early-bred youngsters come from parents that are to be raced then the parents may start their moult too early. The solution is to prevent the laying of the second round of eggs by parting the birds. It sounds simple to remove the cocks – one would do in any case with Widowhood cocks later in the season – and to let the hens get on with the feeding of the youngsters. In practice it calls for great skill and judgement and novices should try it only with great caution. The hens will normally cope with the double task of feeding, but sometimes the babies may have to be hand fed pea by pea by the human substitute. This can be the best time for pellet feeds, making it easier for the hens and the babies. Needless to say this system works both with birds raced Widowhood and those on the Natural system.

In both cases the birds are separated in mid-January and are returned to each other in mid or late March in the case of Natural racers or perhaps even earlier in the case of Widowers. The exact date depends on the first race. In the case of Natural birds they will be raced sitting the second round of eggs whereas the Widowers will be left together for a few days before being separated for a second time in order to race. The season's campaign needs to be planned with a calendar with some precision and if you, like me, have an indifferent memory, you need last year's programme and notes to get this year's right. No wonder the Belgian professionals are full time! Of course there is a downside. Are January-reared youngsters worse than those reared in March, particularly when many argue that the best are reared in May, June or even later? The January rearers usually have a closed loft with some form of heating and many have time clocks to extend the hours of natural daylight. I am talking about northern Europe, where the shortest day is in December and the coldest months usually January or February. Is it worth the trouble? This question brings us full circle because with Widowhood it becomes much more important to race young hens than under the Natural system.

Turning to the Natural system itself, at the risk of repeating myself, there is a natural sequence that needs to be followed to win on this system and although Widowhood is very popular at the moment the merits of Natural racing cannot be ignored and as explained earlier you need to understand Natural racing even if you intend to race Widowhood from the start. Perhaps now is the moment to touch on the use of cortisone by some fanciers. At the time of writing, it is not a banned substance but I have no doubt that in due course it will be. I am not an expert in its use and don't expect I ever will be but cortisone is used not so much to speed the birds up but to slow the moult down. It is administered from the bottle with an eye-dropper into the eye or less surprisingly into the mouth. It can be obtained only from a vet. As it is not illegal, there are plenty of vets who will prescribe in Belgium but, as far as I know, none in the UK.

For all I know it may work. Some professional YB racers in Belgium are widely suspected of using it but the ordinary genuine amateur is not going to find much use for it. The prize money in some of these Belgian races is large, very large sometimes, but to win with cortisone you must sacrifice any hope of OB success. Indeed it has been suggested even that the YBs after racing are either sold or culled. This is not my idea of pigeon racing. Many years ago when I was a young man there was talk of using drugs to make pigeons race faster. The favourite one in those days was the deadly poison, arsenic, which if given in minute doses would speed the heart up and improve performance. I didn't believe it then and don't now. In the first place, the dose would have to be measured with scientific micro-accuracy and, in the second place, the time of liberation, even the day of liberation in some cases, cannot be fixed so that the stimulation could come at the wrong time.

I wish I knew more about the use of cortisone because the suggestion that it only affects the moult does not seem to fall in line with what we know about human athletes. In humans it is called steroid abuse and any reader of daily papers should be able to name a runner or a swimmer or a weightlifter who has been found guilty of drug taking. Steroids is the short name for corticosteroids, part of the cortisone family. The same family also includes the inhalers used by asthma sufferers to increase the size of their breathing passages. They are so widely prescribed for ordinary medication that many athletes claimed they failed drugs tests because of the medication used. That argument does not belong here; the reason for introducing it is that if corticosteroids improve human athletes surely they would also improve pigeon athletes. There is a price to pay – not just that of getting caught but the long-term side-effects that are already being seen in human drug abusers. I don't want to win using such methods but I don't want to be beaten by cheats.

I am also worried that outside northern Europe and America drug use may prove more widespread. I am thinking of Australia but the same probably applies to South Africa among the English-speaking countries. As I understand it, in Australia very few pigeons survive beyond the yearling stage because of predation by hawks. And because fanciers do not race in the hottest months, their young birds are much more mature than ours. As a result, large YB teams are bred from stock birds each year. Under these conditions the potential for doping must be far greater. However this assessment is based on only limited information.

It is obvious from what has been written that this sort of problem only becomes really important if winning YB races is the main object. This is not what I believe in. To me the target is always the long distance OB races, 500 miles (800 km) on the day, and for this reason the extremes of YB racing are not for me because they cannot be reconciled with 500-mile (800-km) racing. Another reason why some fanciers breed YBs in January is so that when they start to race, the YBs can be already paired up. Quite a few fanciers will let the young cocks and young hens run together and if they pair up so much the better. Particularly useful for this are the deep box perches. The usual ones are only about 3 in (8 cm) deep but these deep ones are 12 in (30 cm) or more and will encourage the more mature young birds to pair together. They may go to nest, even lay and hatch so that you can have young birds bred from young birds. I don't like this. It seems to me burning the candle at both ends. An older form of this is to pair young cocks to yearling hens. Even if late-bred hens from the previous year are used it does not put too much strain on the hens and no damage will be done except in so far as it is paying too much attention to a side show rather than the main event.

THREE

Food and Feeding

Carbohydrates – Fats – Proteins – Fat Soluble Vitamins – Water
Soluble Vitamins – Animal Protein Factor – Minerals – Legumes and
Cereals – Maple Peas – Tic Beans – Wheat and Barley – Maize – Oil
Seeds – Rice – Total and Useful Analysis – Pellet Feeding – Cafeteria
Feeding – Acorns and Beech Mast – Water – Grit – Trapping – Hopper
Feeding – Exercise – Mid-Week Tosses – Lazy Birds

The importance of choosing the right food and giving it to the birds in the right way cannot be overestimated, and the fancier who solves the feeding problem is well on his way to success. Food is required by pigeons in two main forms; the first is carbohydrates, containing carbon, hydrogen and oxygen. The second form is proteins where the same three elements are combined in a different way with nitrogen. The carbohydrates are the energy-giving foods and the proteins are the body-building foods. A typical carbohydrate in the human diet is bread and a typical protein, meat.

Carbohydrates

The carbohydrates, as they are digested, are turned from their starches into forms of sugar which can be absorbed into the bloodstream. In the bloodstream they pass all round the body to give energy to the muscles and organs of the bird. The digestive process by which these starches are broken down is comparatively simple. If the bird is overfed on carbohydrates, like human beings who eat too much starchy foods, the bird will put on weight and after a time get fat.

Up to a point this will not hurt a bird since a small amount of fat provides a reserve supply of energy. It is not quite certain in what order the process is carried out but roughly a bird obtains energy by first of all drawing on the sugars in the blood in the muscles and the liver. It then draws on the sugar from the blood in other parts of the body, including newly digested sugar.

Examination of the crops of birds seems to indicate that food is not passed from the crop into the digestive organs while a bird is racing, but digestion of the food already inside continues while it is in the air. After the bird has exhausted the sugar in the blood it begins drawing on its reserve of fat. Even in a bird that handles quite hard to the touch, these reserves can be considerable and the loss of weight of a bird during a long hard race is due partly to this re-absorption of fat into the bloodstream. As birds have been known to lose over an ounce (28g) during a race, or about 7.5 per cent of their weight, it is obvious that this is a valuable reserve store of food and a bird that has a little extra flesh on it, providing it is not so fat as to be out of condition, will be a useful bird, particularly in a hard race.

Fats

Carbohydrates and fats are somewhat similar in their composition and the fat contained in pigeon foods is another useful energy food. Fats are, of course, the same to the chemist as oils and the names are used according to whether the substance is solid or liquid in its normal form. Fats are, however, a much more concentrated form of energy than carbohydrates and it is reckoned that the concentration is just over two to one. Therefore, the fat contained in a pigeon's food should be considered in conjunction with its carbohydrates.

Proteins

The protein foods are the nitrogenous compounds in the food. They are also known as the albumenoids, and the proportion of them to the carbohydrates is occasionally known as the albumenoid ratio. Proteins, when they are digested, instead of forming sugars, form amino-acids which are transported like the sugars through the body by the blood. These amino-acids form the basis of the bone, feather and muscle of the bird. No less than 25 per cent of the protein that a bird eats is used in the making of feathers. It is therefore obvious that during the growth of a squeaker and during the moult any shortage of protein will show itself in poorly developed birds. In addition, protein is the food that replaces worn-out tissue in the bird's body. The strain of a long hard race not only uses up energy supplies, but also causes a certain amount of the muscle tissue to be worn out. This must be replaced and for muscle replacement, as for muscle building, protein is essential. From this it can be seen that although carbohydrates have their greatest importance before a race, after the race proteins are more necessary.

Proteins, carbohydrates and fats form by far the greater bulk of the pigeon's food, but in addition there are other items of food that a bird must eat if it is to keep in perfect health. These are the vitamins and the minerals. The vitamins are needed only in microscopic amounts. It is estimated that only five parts in a thousand million of food are all that are required of one vitamin. Vitamins are divided into two main groups: those which are soluble in fat and which the bird is able to store in its body for an appreciable length of time, and the water soluble vitamins which cannot normally be stored internally.

Fat Soluble Vitamins

The main fat soluble vitamins are A, D and E. Vitamin A is one of the most important vitamins in that if there is a shortage, the bird will be more liable to disease, and growth will be retarded. It is normally found in only just sufficient quantities in maple peas and tic beans and is absent completely from all cereal grains except yellow maize. In the green leaves of cabbages and lettuces it is present in considerable quantities. If birds are not given green stuffs in the loft they will quite often go out and pick at blades of grass in order to get it. It is also available commercially in cod liver oil.

Vitamin D is also found in cod liver oil but is obtained naturally by the action of sunlight on the bird's body. Normally the bird is able to manufacture sufficient internally for its own use, but any deficiency, should one arise, would cause malformation of the bones. In human beings, where greater proportionate amounts of vitamin D are required, a deficiency gives rise to rickets. Vitamin A and vitamin D act together rather curiously since if extra large quantities of vitamin D are administered then it will counteract the vitamin A. In addition to assisting in proper bone formation, vitamin D is necessary in forming the shells of eggs and by way of these shells the bone structure of embryo youngsters.

Vitamin E is contained in abundance in maize and all other cereals but not in peas and beans, and in addition is usually to be found in the small seeds. As we have seen, it is this vitamin that has an important effect on fertilization. In a normal pigeon's diet it should not be short, but it is available in a concentrated form as wheat germ oil. The feeding of any cereal will normally provide adequate amounts. Another fat soluble vitamin is vitamin K, which is important for clotting the blood. Sufficient quantities are available in peas and beans for normal use.

Water Soluble Vitamins

The water soluble group of vitamins is far less simple and consists of vitamin C and the B complex, as it is called. Vitamin C, as far as is known, is not necessary in the diet of pigeons and any amounts that might be needed could be supplied by some internal process of the bird. The B complex is the name given to a whole group of vitamins which years ago were thought to be all the one vitamin B. Today the B complex has been split up into many parts such as vitamin B_1 and vitamin B_{12}. Vitamin B_2 was at one time called vitamin G but is now usually known by its name riboflavin. Since this vitamin is contained in all legumes and in cereals, it is scarcely possible for a bird to be short of it, but if a bird is artificially given a diet lacking in this, then there is considerable increase in the number of youngsters that fail to hatch out. A rich source of this vitamin is brewer's yeast or the synthetic riboflavin.

Animal Protein Factor

Vitamin B_{12} is one of the important parts of the animal protein factor. Little or nothing is known about the influence of this animal protein factor on pigeons but in the poultry industry it is of very considerable importance. Although it is not a food in itself, the animal protein factor helps change protein foods into amino-acids that can be absorbed in the bloodstream. Certain forms of protein will not be digested unless minute amounts of animal protein factor are available. For this reason, if what is true for chickens is true for pigeons, then it should be included in the diet. As the name suggests, the animal protein factor is contained only in proteins of animal origin and not in peas, beans, maize, etc.

It would seem then that if full value is to be extracted from the normal pigeon foods then some small amounts of animal protein should be added. As the animal protein factor (vitamin B_{12}) is a water-soluble vitamin, this should form a regular part of the bird's diet and one specialist in poultry breeding who has studied pigeon feeding has said that in his opinion pigeons should receive some animal protein every day. It can be argued quite reasonably that pigeons have been flying 500 to 600 miles (800 or 900 km) regularly and have done it successfully on vegetable proteins only.

However, it is interesting that most of the recipes for 'speed cakes' and other aids to success contain small quantities of meat extract such as Bovril or products derived from fish, both sources of animal protein. In the same way some birds often go down into the garden and pick up snails and insects, other sources of animal proteins.

I am sometimes asked for a recipe for speed cake. There are hundreds of variations but the following are tried and tested old time formulae:

Cheetham's Wartime Formula
Baked bread crusts
2lb (1kg) lentils
2lb (1kg) groats baked again
1lb (500g) jar of cod liver oil and malt
3 tablespoonfuls honey
Dried off in oven and mixed with a little flour.

Pearson's Old Formula
2lbs (1kg) each of rice, groats, hemp, canary, rape, millet
1 cup sherry
2 tablespoons Parrish's food
2 tablespoonfuls sugar
12 eggs
Dried in the sun or in a very slow oven.
Dose: One full egg cup per pair of birds.

Minerals

The final essential constituents are the minerals. These are again required in only small amounts. If the birds overeat then it is possible that they will have a harmful effect but, unless they are fed mixed with another food, this is not likely to happen. The grit a pigeon eats is normally a granite or a flint grit and is used for the grinding of food in the gizzard. The softer grit is the limestone grit and parts of this are absorbed into the body. Limestone is a form of calcium carbonate and the calcium from this is not only the basic material from which the egg-shell is made but is also one of the principal constituents of the bones. If there should be a serious shortage of calcium then soft-shelled eggs will be produced and if any nestlings are hatched, they will be poorly boned. Another mineral needed by the bones for adequate development is phosphorus but this is required in such minute quantities that it is easily obtainable from its natural foods. The same applies to manganese, yet another mineral which has a part in building sound bones.

There are a lot of companies spending a lot of money advertising their products in *The Racing Pigeon* and other papers. These products consist of minerals, supplements and additives. Obviously many fanciers find that these products work for them because they spend money, often a lot of money, on them. Part of my job was, and still is, interviewing winning fanciers and I would be stupid if I did not listen to what they say. I am talking about winners, big winners, and many, indeed most, use these commercial supplements. In fairness I must say that a few do not and yet they still win. It is a minefield but I hope I can give fanciers, particularly beginners, some idea of what to look for.

There are traditional products of varying degrees of usefulness. I believe, for example, in the effectiveness of speed cake. It may be a century old but it seems to me to have some sort of scientific basis (see vitamin B_{12}) and if it were no good I would have expected its use to have died out years ago. I also believe in using garlic. I know that there has been a boom in herbal remedies in recent years, and that there are dozens of herbal remedies that are not only over-priced but about which claims are made that only the overcredulous would accept. The claim for herbal 'cures' needs as much close examination as the pharmacist's chemical cures, but they do not get it, and so are even more suspect than the chemicals. However, I take garlic myself and give it to my pigeons because it seems to me that there is reasonable evidence to believe in the claims made for it. Other additives I am not so sure about. I knew a winning fancier who believed in iron. His system was to put a large nail in the drinking fountain. It was left there to get rustier and rustier and nothing I could do would persuade him that the pigeons could not absorb iron in this form. He was winning, so perhaps he was right but for the wrong reasons.

I am not qualified to be able to separate rival claims for products but what I can give is some ideas for making up your own mind. The first is try and find out something about the company. My simple test is to find out if they have a resident vet and a resident pigeon fancier. Resident in this case does not mean that they actually live on the premises but only that the company has frequent and easy access to these experts for each is dependent on the other for successful product. Let me give an example from the 1994 British National (Old Comrades) Show. The judges for this show are selected entirely on merit. Two friends who were both successful fanciers, one Belgian and one from Yorkshire, had met because of their work for the huge international chemical company Hoechst. Their talk was of the new 20 million pound sterling company to be built in Shanghai. This is big business in which pigeons do not figure. Also at the show was a small trade stand from Hoechst selling and publicizing their worming product. Needless to say, neither person knew of the existence of the other. The trade stand was selling capsules for the treatment of worms in pigeons. There can be no doubt that the product had been scientifically proven to be effective but if the fanciers had been consulted would they have recommended capsules? Capsules must be administered individually to each pigeon and even if you have only forty birds this is a lot of work. Will such a product compete with its rivals that are mixed into drinking water? The scientists may ask for exact dosages but if the broad blanket approach works, and it does seem to, fanciers will need a lot of convincing to invest in the individual capsule method. I have given this example in a little detail, first because it is a specific case that I witnessed and, second, it illustrates the point I was trying to make of the essential linking of scientist and fancier.

A number of the companies who advertise remedies include a phone number for a help-line. This is encouraging because it means that the firm concerned hopes to be able to answer the queries it is bound to receive. This is actually a two-way system because by talking to fanciers the manufacturers can discover what their real concerns are and, who knows, maybe even devise new products. I mention this to encourage those who might be inhibited from using the advice lines. They are an important link between scientist and fancier and should be used as much as possible because the scientists need the feedback from fanciers as well. This feedback is important because it can lead to new and improved products.

I think I have given a few clues about how to look at the

advertisements for the various products. There is more. It is easy to make magical claims for products; less easy to prove these claims. The columns of *The Racing Pigeon* have always been open to advertisers who want to give details of the research that has been done by their company towards the creation of that product. If they want to go into even greater scientific detail, then the monthly *Racing Pigeon Pictorial* is more appropriate. Either way, in the absence of a scientific journal, this is a way to publish the results of any scientific research. The opportunity is not used as much as I would like but it is there and I think it is very important.

Every manufacturer will, naturally, claim his product is the best but how is a fancier to know? My suggestions are:

(1) Look at the parent company. Does it have a track record in pharmaceutical research?

(2) Look at the operating company. Does it have a track record of familiarity with the world of pigeon racing?

(3) Try to identify people in the company who are involved with pigeons and the practicalities in the sport.

All this should be done before you look in detail at the products. Do they look as though they are the result of proper pharmaceutical or nutritional research? Do they look as though they have been put together by someone who doesn't know one end of a pigeon from another?

I am very fond of using the *Squills* barometer. For many years I edited a year book that bears the pen name of my grandfather, the first in the line of writers on pigeons. *Squills* will be 100 years old in 1998. Over all those years the basic formula has been the same for the editorial pages: ask the winning fanciers of that year to describe in their own words how they did it. The articles are written for the benefit of novices and obviously are a treasure trove of winning methods. My barometer consists of going through the articles after publication and just counting, in this case, those who mention minerals or additives or similar. Not all mention what they use and I take that to mean they don't attach great importance to them, a neutral rather than a negative position. Some are negative and these are recorded too. The result is the nearest to a true record that you are likely to get.

Total number of champions recorded __

Number reporting use of minerals __

Number reporting use of additives __

Number of these using both __

Number not recording either __

Number disclaiming additives __

Percentage expressing preference for additives __

Legumes and Cereals

The two main types of foods on which pigeons are fed in this country are the legumes (peas and beans) and the cereals (wheat, maize, barley, etc.). When I first started racing we always fed the birds on Tasmanian or New Zealand maple peas. Since then, English maple peas have been very much improved but, more than ever before, the small English-grown tic bean has become the main feed. These are the two main legumes; the other is vetches or tares. Legumes are all roughly similar in their composition, although the analysis does vary slightly from district to district and from year to year. In the table of feeding substances on page 73 a protein figure is given for tic beans. This is the figure I have obtained from the scientists but my own experience leads me to think that it is possibly a little on the low side. After one very dry harvest I took samples from Essex farms that were growing tic beans and had them analysed. There were several surprising results. The first was that, even on one farm, from one field to another there could be a difference of as much as 2 per cent in the protein values. Second, when we screened the beans, only analysing the smaller ones, we discovered that these had a higher protein content than the larger beans. The scientists told me that this was because most of the protein is concentrated round the outer skin. Third, and most importantly, the protein content could be over 5 per cent greater than the scientists' figure. This is one of the reasons why the price of English-grown beans continues to escalate. They can be ground up and mixed with meats in cheap sausages, they can form food for vegetarians, and nowadays they are even made into synthetic steaks. They are a wonderful source of food and humans are only just catching on!

In order to work properly a racing car needs fuel and maintenance performed by a mechanic. A racing pigeon uses carbohydrates for its 'fuel' and proteins for its maintenance. The cereals are not so rich in body-building proteins but they have a larger proportion of carbohydrates. In addition, the fats contained in the cereals are important sources of energy, particularly as these fats are more concentrated fuels than the carbohydrates. In this respect oats and maize are far more important than wheat and barley although otherwise there is little difference between the cereals.

A diet has to be found such that the necessary proportions of carbohydrates, proteins and fats are supplied to the bird. The fancier unfortunately has other things to consider as well as the best diet, in particular his pocket and the likes and dislikes of the birds.

Beans are not always a popular food with birds but maize usually is and some fanciers make sure that birds eat what they think is the right proportion by giving them beans first and then following with maize. This is a little bit too fiddly for me. I restrict the quantity so that the birds have to clear up the food at each meal time.

In recent years there has been a swing away from the almost traditionally high percentage of beans and peas in the diet of British pigeons. This is partly because of the increasing influence of Belgian diets but also the influence of price. As I will be discussing later, this concerns the so-called depurative or weak mixtures and the new range of Belgian foods with a whole new range of ingredients in small quantities. Very few diets will contain more than 50 per cent of high protein foods. In many it is down to about a third: in other words the racing mixture of today is more like the winter mixture of years gone by. There are many corn suppliers who make up a winter mixture which contains beans, maize, barley and a small proportion of maple peas. The racing mixture will consist largely of beans with additions of maize and sometimes peas. A fancier can experiment by making his own mixture to get a good balance of body-building and energy foods. One successful fancier I know uses a commercial racing mixture to win but in order to minimize the variation actually buys three different racing mixtures and mixes them together.

Maple Peas

In recent years, with the increase in English arable land devoted to tic beans, one of the greatest changes has been the decline in the maple pea, but there are many fanciers who still prefer the maple pea even though it costs more. In buying maple peas there should never be any hesitation in buying the orange-brown ones. These should not be wrinkled, black or too big. New Zealand peas are dried in the sun and stored for some time before they reach this country. English maple peas are darker in colour since they are not as a rule sun-dried, but have been dried artificially in kilns. To my mind these peas are marginally not as good as the naturally dried ones. The peas should also have a chance to mature after ripening. It is difficult to say what goes on inside or outside the pea whilst it is maturing, for there is little difference to be seen between a matured and unmatured pea, but maturing certainly makes them more palatable to the birds, which are less liable to have loose droppings.

Tic Beans

There has been enormous progress made by the agricultural botanists in producing smaller and more quickly ripening tic beans, and nowadays they are not so large as they used to be. They are, however, still larger than the maple pea, and for that reason are sometimes not chosen as feed for youngsters. They can be a little darker in colour than the best-quality maples and English beans are as good as any of the imported ones. Indeed many of the British tic beans are exported to Belgium only to return as a constituent of more expensive Belgian mixtures. I will go into the constituents of these mixtures later, after first considering the basics. The question of darkness in a pea or bean is dependent on the amount of water at the time of the harvest. If it is a wet harvest then the beans will be darker whether they have been artificially dried or not, but if it is a very dry harvest then the beans will be much lighter and probably not need to go through a drying process at all.

The selection of beans has now become even more important since I suspect that a great deal of the corn used by pigeon fanciers now comes direct from the farm. Although we talk about polishing and cleaning commercial corn, many fanciers go for the cheaper grains available from the farmer which are often barely clean from the dirt of the fields. It is always wise to deal with a firm where you are absolutely confident you will get good quality foods. Most of these will send a sample of their foods on request. If when the goods turn up they are not as the sample, then they should be sent back straight away with a very sharp note, but this very rarely happens. One of the dangers is that the peas may be riddled by worm or they may be soiled by water or mould so as to be unsuitable; these should be rejected immediately. Some varieties of beans are very light in colour but these dun beans seem to be just as good for feeding.

Wheat and Barley

In wheat the principal thing to look for is ripeness. A pappy seed or a green seed should not be considered. Barley is a grain favoured by some and condemned by others. It is of high carbohydrate and low protein value and thus while of use in winter should not be used alone during the moult. The malted barley that has been treated so as to reduce the starch inside is favoured by many users. Barley in the last decade has come in for a new lease of life. In the old days fanciers like my grandfather and Dr W. E. Barker, who wrote one of the best books on pigeon racing, always recommended barley to quieten the birds down after racing was finished and the moult had been completed. I don't think there was then a winter mixture sold in this country that did not contain barley. One of the reasons was that although it was nutritious it was lower in protein than peas and beans; also it was thought to be less stimulating to the birds. Possibly this last idea arose because pigeons were never enthusiastic about barley and in a mixture it would be the last grain to be eaten.

The modern concept among Widowhood fliers involves 'building them up' followed by 'breaking them down'. Immediately after the race the birds are broken down by being given a weak, or depurative, mixture of grains. Don't ask me what depurative means. I don't know other than that is what the Belgians call it. This weak mixture contains a fair proportion of barley and sometimes some wheat. It is given to the birds usually Monday and Tuesday and after that the mixture is changed to build them up for the next race, cutting out the barley and increasing the strength of the mixture with peas, beans or maize.

There is a further complication in that Belgian mixtures no longer consist of three, four or five grains but regularly contain ten or twelve. Some of these extras are small seeds we are familiar with: red rape, black rape, dari, linseed, etc. Others are different varieties of field peas, Canadian white peas, hard green peas for example, with different varieties of beans, like Mung beans grown in China for bean shoots. The only thing these seeds have in common is that they are expensive. Mung beans are four or more times as expensive as tic beans. Are they four times better? I am not convinced. There may be some point in building up and breaking down but it won't make winners out of losers. I think the use of exotic mixtures is even more marginal, for the food values of these rare additions are very close to the conventional ones they replace. Look at the food tables to see what I mean (page 73).

Maize

Maize is usually very popular with the birds and of this grain there are two main varieties. One is a small round tablet-sized grain and the other a large flat grain of which the edges can be quite sharp. This larger grain is not recommended as it is too big for the bird to eat comfortably. Yellow, not white, maize should be given as only this contains vitamin A. Maize has a reputation of harbouring disease and maize-fed birds are often said to be more prone to sickness, particularly canker. I feel sure myself that it is not the grain that is at fault but rather the large size which makes the bird open its beak very wide and perhaps causes a slight split in the skin at the corners of the mouth. It is this split that becomes a centre of disease. However, this is only a guess since it has never been proved to me that disease is more common in maize-fed birds although it is an often-repeated accusation. Maize is the most valuable of the energy foods and the best of the cereals.

Oil Seeds

Many of the seeds used in these mixtures are called oil seeds for the obvious reason, as the table shows, that they are very rich in oils. Linseed has been used for years because it is not only a very rich food but also because it is believed to improve the plumage. Others are thought to have aphrodisiac qualities but I have never seen any scientific research to support any of these claims. What is certainly true is that the birds welcome these changes in their diet. Many of these oil seeds are used in the so-called trapping mixtures, tit-bits that are given to birds on arrival from a race as a reward to be looked forward to. For this their food value is almost irrelevant.

Rice

Another food well worth mentioning is rice. For domestic use this is usually sold polished, but in this way much of the valuable properties are lost. Therefore rice on which the husk remains (undercorticated rice) should be fed whenever possible. O. I. Wood, the famous fancier from Ilkley, Yorkshire, was a great believer in rice. It has the highest carbohydrate value (78 per cent) of any of the pigeon foods and he regularly used this just before the bird went to a race and just after. I just cannot understand why this food has fallen out of favour. I know seventy years have passed since Ossie Wood was wiping the floor with the opposition, but that does not alter the fact that he was winning out of turn and that rice is a superb source of energy. Whatever the price may have been years ago, it does not bear comparison with the modern mixtures. When non-fanciers report a found pigeon I always suggest they feed rice rather than the very expensive wild bird mixtures they might have in the house and it is a good food for quick recovery.

It is, of course, worth noting that while carbohydrates are the energy food, an increase in them does not necessarily mean greater speed, since the pigeon may easily have sufficient carbohydrates as it is. A deficiency continued over a time can mean that the pigeon will not fly as fast as it is able. Indeed when a pigeon is flying a hard race the protein consumption goes up at a much higher rate than the carbohydrate consumption.

Total and Useful Analysis

In considering the differences between various foods we have talked about them in general terms. The table shows in detail the composition of these various feeding stuffs. It is important to notice the differences between the column showing the chemical quantity obtained by analysis of proteins, carbohydrates and fats in the food, and the other column showing the amount of these proteins, etc., that can actually be used. This is a very important difference, and in planning a diet the actual amounts of protein that can be digested are the important factors. Unfortunately it is not so easy to be accurate with these figures since the amount that is digested can vary according to the other foods that a bird is fed. A minor point worth noting is that linseed, as well as having a high fat content, is very high in proteins and it is this as much as anything that gives it its value during the moult.

Pellet Feeding

If the question of feeding were as easy as just giving the right amounts of proteins, carbohydrates and other foods then it would be perfectly simple to make up a food out of other substances and feed it to the birds. Attempts have been made to cut the cost of feeding by using various substances frequently used in poultry mashes.

It can be seen from the table overleaf that there are some excellent foods available, and that fed in the right proportions they would give a pigeon adequate nourishment. One Colchester fancier has raced successfully for years on poultry pellets and his birds even come down for ordinary wet mash if a bit hungry.

The real difficulty in feeding these pellets arises from what is known as the metabolic rate. This can be thought of, for present purposes, as the speed of digestion. In other words, it is the time it takes a bird to extract the energy from the food it eats and to get this energy into the bloodstream to all parts of the body. Since the speed of digestion in chickens is faster than in pigeons, poultry pellets will pass through quickly when they are given to pigeons, and the bird may not extract all the food from them.

For this reason pigeon pellets must be made to digest slower than poultry pellets and those on the market at the moment are considerably harder than poultry mash pellets. In the preparation of pigeon pellets the problem of getting them hard enough has been solved largely by reducing the amount of fibre in the pellets. The ingredients and analysis of two of the first commercial pellet mixtures are as follows:

Purina Pigeon Chow Checkers (USA). Ingredients: yellow corn, Kafir corn, soya bean oil meal, dried whey, dehydrated alfalfa, wheat middlings, calcium carbonate, trace minerals, vitamins – B_{12} riboflavin, A and D, niacin, calcium panothenate. Analysis (digestible): protein 13.5 per cent, carbohydrates 56 per cent, fats 2.5 per cent, fibre 4 per cent, ash 6.5 per cent.

Hormofeed No. 1. Ingredients: peas, beans, maize, barley, alfalfa meal, wheat middlings, rice meal, linseed, fish meal, meat and bone meal, vitamins A, D, B_1, B_2, B_{12}, and Hormoform (an opotherapeutic food supplement). Analysis: protein 17.5 per cent, carbohydrates 49.5 per cent, fats 3.5 per cent, fibre 3.5 per cent.

TABLE OF COMPARATIVE VALUES OF FOOD

Feeding-stuff	Dry Matter	Protein		Carbohydrate		Oil		Fibre
		Crude	Digestible	Soluble	Digestible	Ether extract	Digestible	
	%	%	%	%	%	%	%	%
Legumes								
Maple Peas	86	22.5	19.4 (16.1)	53.7	49.9	1.6	1	5.4
Tic Beans ⎫ Dun beans ⎬	85.7	25.4	20.1 (19.3)	48.25	44.1	1.5	1.2	7.1
Tares ⎫ Vetches ⎬	86.7	26	22.9 (20.0)	49.8	45.8	1.7	1.5	6.0
Soya beans	90.0	33.2	29.5	30.5	20.8	17.5	15.8	4.1
Cereals								
Maize (and maize meal)	87.0	9.9	7.1	69.2	65.7	4.4	3.9	2.2
Wheat (strong varieties)	87.9	13.0	11.4	69.6	64.0	2.2	1.2	1.6
Wheat (weak varieties)	88.5	9.5	8.4	73.8	67.8	2.0	1.2	1.7
Barley (and barley meal)	85.1	8.6	6.5	67.9	62.2	1.5	1.2	4.5
Oats	86.7	10.3	8.0	58.2	44.8	4.8	4.0	10.3
Dari (Kafir corn)(Sorghum grain)	88.9	9.6	7.7	71.2	60.5	3.8	3.0	1.9
Buckwheat	85.9	11.3	8.5	54.8	42.3	2.6	1.9	14.4
Rice (polished)	87.4	6.7	5.8	78.0	75.8	0.4	0.2	1.5
Oil Seeds								
Linseed	92.8	24.2	19.4	22.9	18.3	36.5	34.7	5.5
Hemp	91.1	18.2	13.7	21.1	16.8	32.6	29.3	15.0
Sunflower seed	92.5	14.2	12.8	14.5	10.3	32.3	30.7	21.1

(continued overleaf)

Feeding-stuff	Dry Matter	Protein		Carbohydrate		Oil		Fibre
		Crude	Digestible	Soluble	Digestible	Ether extract	Digestible	
	%	%	%	%	%	%	%	%
Ingredients for Pellets								
Wheat middlings (fine)	87.3	15.7	13.2	64.0	52.0	3.4	3.0	1.8
Wheat middlings (coarse)	86.5	16.4	13.8	56.2	45.5	5.0	4.3	5.3
Wheat pollards	86.7	14.3	11.6	55.6	44.5	4.8	4.0	7.7
Wheat bran	86.4	13.5	10.6	53.0	38.0	3.9	2.8	10.6
Maize (germ meal)	89.3	13.0	10.4	55.1	45.8	13.5	12.8	4.1
Maize gluten feed	89.6	23.5	20.0	56.7	49.3	3.4	2.7	3.5
Maize gluten meal	90.9	35.5	30.6	47.5	42.6	4.7	4.4	2.1
Oat feed (high grade)	91.0	10.3	—	53.7	—	5.0	—	17.5
Oat feed (ordinary)	92.0	5.5	—	52.5	—	2.4	—	27.6
Oat husks	94.0	2.0	—	54.0	19.4	1.0	0.4	33.0
Brewers' grain, dried	89.7	18.3	13.0	45.9	27.6	6.4	5.6	15.2
Soya bean meal (extracted)	88.7	44.7	40.3	31.9	24.7	1.5	1.4	5.1
Palm kernel meal (extracted)	90.0	19.0	17.1	49.0	43.5	2.0	1.9	16.0
Ground nut cake meal (decorticated)	89.7	46.8	37.6	23.2	18.6	7.5	6.7	6.4
Sunflower cake	90.4	37.4	33.6	20.4	14.6	13.8	12.2	12.1
Linseed cake meal (English)	88.8	29.5	25.3	35.5	28.5	9.5	8.7	9.1
Linseed cake (foreign)	89.0	32.3	27.8	32.2	25.8	9.9	9.1	8.7
Linseed meal (extracted)	88.2	35.7	30.8	33.9	27.2	3.1	2.8	9.0
Biscuit crumbs	90.6	13.9	12.5	74.6	68.5	0.7	0.6	0.3

Vegetables and Wild Seeds								
Acorns (fresh)	50.0	3.3	2.7	36.3	36.6	2.4	1.9	6.8
Acorns (dried)	85.0	5.7	4.6	61.6	55.5	4.1	3.3	11.6
Beech mast				(Approximately as Linseed)				
Lucerne (in flower)	24.0	3.9	2.7	9.2	5.7	0.8	0.4	7.8
Vegetables and Wild Seeds (cont'd)								
Alfalfa (lucerne) meal	89.3	14.6	9.2	33.5	22.0	1.8	0.9	31.0
Cabbage	11.0	1.5	1.1	5.9	4.6	0.4	0.2	2.0
Kale	14.8	2.5	1.8	8.7	7.0	0.3	0.2	1.7
Potatoes	23.8	2.1	1.1	19.7	17.7	0.1	–	0.9
Potato peelings (hand)	21.2	2.1	1.6	17.0	16.4	0.1	–	0.7
Yeast (dried)	95.7	48.5	41.6	35.5	29.2	0.5	0.2	0.5
Meat and Fish Products								
Fish meal	87.0	55.6	50.0	2.1	–	4.4	4.2	–
Meat meal	89.2	72.2	55.7	–	–	13.2	12.5	–
Meat and bone meal	90.8	50.5	39.0	–	–	10.0	9.4	–
Blood meal	86.0	81.0	72.7	1.5	–	0.8	0.8	–
Milk, fresh (whole)	12.8	3.4	3.2	4.8	4.8	3.9	3.9	–
Milk (whole dried)	95.8	25.5	24.0	37.4	37.4	26.5	26.5	–
Milk (separated)	9.4	3.5	3.3	5.0	5.0	0.1	0.1	–
Milk (skimmed dried)	89.8	32.8	29.6	48.8	48.8	0.3	0.3	–
Buttermilk (dried)	90.0	42.3	34.5	24.3	24.3	11.2	11.2	–
Whey (dried)	92.2	12.6	–	70.5	–	1.4	–	–

These were the pioneer formulations of pellet feeds. Today there are many more, too many to analyse in detail, but I think one range produced by a specialist firm deserves some attention. This is how it is described with their suggestions for use.

Spillers Top Flight Moulting – Pre-Breeder Pellets are specifically designed for birds from moulting through to breeding. High in the amino acid methionine, they will ensure healthy and rapid feather re-growth. The combination of energy protein and minerals like calcium will build up body reserves that will be needed during the breeding and laying period. Moulting-Pre-Breeding pellets should be introduced when feathers drop and fed until pairing-up when Spillers Breeding Pellets should be introduced. They can be used as a complete feed or mixed with Top Flight Moulting Mix. The pellets are a blend of cereals and proteins and do not contain fish meal or any other animal by-products. Typical analysis: protein 16 per cent, oil 5 per cent, fibre 4.5 per cent.

Spillers Top Flight Breeding Pellets are specifically designed for all breeding birds and young stock. Normally, breeding pellets should be introduced at least 3–4 weeks before eggs are laid (at pairing up). They can be used as complete diet or mixed with Top Flight Breeding Mixture. Up to 40 g per day can be fed throughout the racing season to help overcome mineral/vitamin deficiencies in the birds' diet if feeding all corn diet. No supplementary minerals or vitamins need to be fed. They do not contain fish meal or any other animal by-product. Typical analysis: protein 18 per cent, oil 5 per cent, fibre 4.5 per cent.

The advantages of pellet feeding are first and foremost their lower price and second the ability to feed a balanced and supplemented diet to the birds. On the question of price, at the moment the pellets that are available are at least as cheap as a winter mixture and these pellets have the same protein and carbohydrate value as the best food obtainable, with the addition of vitamins and minerals. The question of the balanced diet is one on which there is now and always will be considerable discussion. Some of the old fanciers will say peas and beans have been good enough for years, how can they be improved upon? Other fanciers realize that in the field of nutrition perhaps there is something they can learn and will be willing to experiment.

Pellet feeding is not without its disadvantages; the principal of these is that the pigeon is an animal of habit and having eaten peas and beans all its life may not take kindly to eating pellets. Indeed some would rather starve than eat them but most, like my own, take to them without trouble and once having taken to them can be fed on peas or pellets as the fancier wishes. There are very few birds which will not take to them if kept hungry enough, and once having eaten them, there is rarely trouble after that. Pellet feeding is comparatively new for pigeon racers. So is the idea that peas and beans might not be sufficient in themselves. My own experience with pellets has been that their great advantage, once you get the birds used to them, is during the breeding season, because when the parents pump back food into babies they are pumping back an easily digested wet mash rather then hard-to-digest corn. Certainly the birds thrive on this and therefore the plan I have adopted does not require slower-digesting pellets. It is also useful to use medicated pellets.

Cafeteria Feeding

The cafeteria system of feeding is a system in which all the different varieties of corn and seed are placed separately in front of the pigeons so that they can help themselves. It is argued that since wild pigeons must be able to select a good wholesome diet to exist at all, tame pigeons will still, if given a free choice, eat the correct foods in the correct amounts. With cafeteria feeding a mixture can be fed without mess whereas if the same mixture is fed in a hopper it is often scattered by the birds over the floor. The whole idea hinges round the question of whether birds will actually choose the proper diet. According to squab producers in the USA they do; but even so, they cannot prepare for an on-coming race as a fancier can. For the racing fancier the cafeteria on its own is not enough.

Acorns and Beech Mast

Among the foods that are commonly eaten by wild pigeons are acorns, and many racing fanciers have used these for feeding their birds. One old-time fancier from Woking fed them throughout the racing season and achieved more than his share of success. During the 1939–45 war they were used by many other fanciers when corn was unobtainable for many and was short for all. Even today acorns have still been used by those who want to cut down their expenses a bit and who have sufficient time and energy to collect them. Fanciers who have been able to enlist the aid of small boys can usually get them collected at a reasonable cost. The analysis of acorns is given in the table and it will be seen that it is very similar in many respects to barley which has often been recommended by old fanciers for a winter food. Even more valuable during the moult is beech mast. This, when it can be obtained, is rich in protein and in oil and is roughly equivalent to linseed. In these times of ecological awareness the idea has its attractions.

Care has to be taken in giving these wild seeds to the birds. They must be collected when they are ripe, stored until they are dried and then broken up before being given to the birds. The husks have very little value as food and if the fancier has time can be removed. If they are fed to the birds with the kernels, then the birds will leave them and they can be removed when the birds are cleaned out. While acorns are being dried they should be spread evenly over a large area. If they are kept together in a bin then there is a danger that they may start working and be unsuitable for food. There are many other types of wild berries and foods that were tried during the war but none has proved very satisfactory as a food. Indeed the top prize winner in one North London club fed his birds for a whole year on toasted wholemeal bread and margarine but such emergency feeding cannot be profitable in the long run.

Water

While a certain amount of variation can occur between fancier and fancier on the foods they recommend, there is no question of any disagreement on the subject of water. Birds must have water before them all the time and it must be pure, fresh water. If the water is given to them out of drinkers, these should be replenished often by emptying and refilling so that they are always nearly full. This is particularly important in hot weather and when parent birds are feeding youngsters, since at this time they use large quantities of water. In addition the drinkers should be cleaned out at least once a day by rubbing round with the fingers and given a more thorough cleaning once a week. In some cases fanciers with either the building skill or a deep pocket have fitted their lofts with running water and this has a lot to recommend it. Although water differs in many parts of the country, being very hard and full of calcium carbonate in many parts of the south, and very soft particularly near the Pennines, it does not appear to make much difference to the birds. When birds are transferred from one district to another there may be some looseness for the first week but as a rule this does not persist.

Water is also important to the birds for their bath, and the birds should be allowed to bathe at least once a week. Outside my loft there is an old sink and when the birds see this being cleaned out preparatory to filling it up with their bath water, they line up on the far side so that they are all ready to jump in. Once in, their pleasure is obvious and they are undoubtedly cleaner when they come out. Even when there is no permanent, fixed bath a large tin should be put out with 2–3 in (6–8 cm) of water in it once a week so that the birds can bathe. A fine sunny day is obviously preferable, and if it is damp and muggy, the birds will not suffer from the lack of a bath. The Sunday morning after a race is usually a good time to give it to them as it helps clean up the racers that have returned.

Grit

As well as food and water there are several other items that the birds need. The importance of grit for the gizzard has been mentioned before. In some parts of the country it is possible to obtain it locally from the seashore, and mix it with crushed oyster shells, but most fanciers buy it commercially from grit suppliers. I am not altogether convinced that naturally produced grit from local sources is desirable, since, if birds can pick up a similar substance outside the loft they may well take to fielding. The dangers of fielding are discussed later but it is worth mentioning here that well-known fanciers have lost birds from poisoning by artificial fertilizers, and it is necessary to mention the dangers of seed dressing, even though the most lethal are now banned by law.

Trapping

Apart from the poison danger, birds given to fielding will become bad trappers, sitting out on arrival at the loft. Bad trapping can be caused in many ways but fielding and open loft are among the most common. In days gone by you needed a good fancier and a good pigeon to get a bird from 500 miles (800 km); today you still need a good fancier and a good pigeon, but there are many more of both and even in the long races the bird that sits out for any length of time at all can lose the race. In the short races it is not minutes but seconds that count, and if the birds do not come straight into the loft, you may be not one place lower, but half way down the list.

Birds must get into the habit of trapping immediately they are called into the loft. They must be taught to trap at once when some signal is given. The signal can be either a whistle, a voice, the rattling of a corn tin, the ringing of a bell or anything which is distinctive and used only to call them in. The surest way they can be taught to come in on this signal is by using it at their feeding time. They must learn that the rattle of the corn tin means that it is feeding time and that they must come in at that moment for their food. If they have food in hoppers before them constantly or if they are able to obtain food from the fields then they will never learn to obey the rattle of a corn tin.

Good trapping can be induced from birth by feeding control but this applies mostly to young birds. With old birds it is the nest, the mate that matters most. The ideal situation is open-door

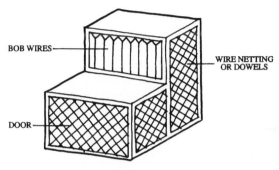

9 THE 'PIANO' TRAP.

trapping. The bird comes out of the sky through the open doors of the loft straight to its nestbox. The fancier follows it and takes the rubber off. In theory no time is lost but many birds returning from a long race are highly strung and more nervous than usual, so the fancier must close the loft doors behind him, losing a few possibly vital seconds.

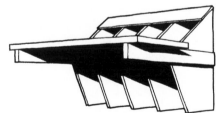

10 THE 'YANKEE' TRAP. THE BOB WIRES ARE TO THE REAR AND RELEASED FROM UNDERNEATH.

The present vogue is for 'Yankee' stall traps. The entrance the bird has used day after day is like the old 'piano' drop hole but the modification is that on race days the bob wires do not open at

81

will and the pigeon is secured in its stall so that the rubber can be removed before the bird actually enters the loft. Some of the high-speed specialists actually sit beneath the stall traps to save those valuable seconds. There are a lot of things in favour of stall traps, not least of which are the facts that the birds do not need to be as tame as those for open-door trapping and that because the rubber ring is there ready to be removed, seconds may be saved. If you are into sprint racing, this is the system for you, but as I keep on saying if you are into long-distance classic racing, it does not have the same importance and may even have some negative aspects. A pigeon, whether raced Natural or Widowhood, returns to its mate, its nestbox, even its owner. If it returns and finds the loft shut up this has got to be discouraging and even the delay caused by the stall trap seems to me to be unwelcoming.

Hopper Feeding

For trapping reasons the food hoppers should not be before the birds continuously. If the fancier has not the time for hand feeding twice a day, the hoppers should be put in only when the birds have been called in from exercise. If the birds are flying twice a day they should be put in after the morning fly and removed half an hour later. They can stay in after the second feed until the evening. This is quite commonly done by many fanciers, but it is also quite common that in winter, the birds are allowed to have the hopper all day. Although exercise is not so important in winter, the birds should still have a fly whenever the weather permits. The hoppers should be put in after their fly and then left until the evening.

In this way the birds hear the calling-in signal practically every day of their lives and it will soon become a habit for them to trap immediately at the fancier's signal. If this habit can be made part of the birds' lives then even if they are not hungry and even if there are things that they would far rather do, they will come in to the rattling of the corn tin. In winter, of course, and quite often in summer the weather can be too bad for the birds to fly and on days like this the hopper can be put in during the morning. Better still, the birds can be made to feed from the hand, and in this way will receive further useful training of a different sort.

Exercise

The demands of young birds and parents feeding youngsters should not be overlooked, for at this time the food eaten and water drunk increases very considerably. In particular, birds feeding youngsters in the nest should never be kept short and this must be considered in deciding whether to feed by hand or from hoppers. The great advantage of hand feeding is in the close contact that the fancier can make with his birds. This will be discussed in more detail later.

In discussing the problem of birds that sit out, it has been mentiond that this can be cured or need never arise if the right steps are taken during the exercise periods. The problem of exercise is one on which opinions are most divided. A frequently heard remark is a fancier saying his birds will not exercise round home. He turns them out for their morning exercise, they fly for a few minutes and then keep trying to pitch on the loft. If they are chased from the loft they will eventually pitch on an inaccessible house-roof. Many and diverse are the methods which have been devised to keep them flying. Renier Gurnay, the Belgian Widowhood expert, used to have a mast by the side of his loft and he kept running a flag up and down this to keep the birds on the move and this is used by many fanciers today. Some fanciers spend many hours shouting and hat-waving to keep the birds going and in this way the desired amount of exercise can be got in during the day. Others say that scaring them is all wrong.

Opinions vary on the time in the air that is necessary. Some say half an hour twice a day is enough; others demand much more and a few others will accept less than this. The sharp difference of opinion on this subject shows how much depends on the individual birds. I made a slow motion film of birds flying around a lóft, and I was surprised to see that quite often the birds were moving their wings very gently indeed. When this was compared with the tremendous effort made during a race, most of the time flying round the loft was not really used exercising, since the birds scarcely exerted themselves.

Mid-Week Tosses

Many fanciers have recognized this in a different way and have stated that their birds do not fly well round the loft and so they give them one or more mid-week tosses in between races. These fanciers have won high honours in Combine and Federation races so the system must work, at least for them. The other school of thought is that if the birds do not show willingness to take their exercise round the loft and enjoy it, then they are not in good health. There is a lot to be said for this argument. The birds should enjoy their daily fly and in good weather should revel in their exercise, but this again is a thing which, while true for some birds and some fanciers, is by no means true for all. The birds undoubtedly need exercise and if they do not take it round the loft they should be given it from the basket.

Basket training is to my mind satisfactory in many ways: first, because it gives the birds increased basket experience, which is always valuable and, second, because it forces the birds to take complete and strenuous exercise rather than the casual flapping round home. Mid-week tosses can, of course, be overdone and the bird's strength reduced for the major contest at the end of the week. In the first races, two 30-mile (48-km) tosses are sufficient, reducing the number (not the distance) above 150 miles (240 km). With Widowhood birds most of the tosses are pre-season when the cocks are still with their hens but there is a need for real Widowhood tosses. In other words, once the cocks have been separated the hens are brought back just as if it were a race. This means the hens are shown the cocks before the toss and will be waiting for their return. With old birds it may be necessary to do this only once to remind them what Widowhood is all about but with yearlings you will be talking about at least 3–4 tosses. They have got to learn before the first race.

The early tosses, when they are still paired, are important and, weather permitting, you should aim for four or five up to 30 miles (48 km). After they have been put on Widowhood, with, it is hoped, improving weather and longer daylight, try for at least two at 30 miles (48 km) and a longer one if possible. This is most important for the yearlings and since the other birds are going to be in the same loft it won't do the older ones any harm.

Basket training, unfortunately, is quite costly in time and petrol and sending them too far and too often is something not many of us can afford. Tosses shorter than about 30 miles (48 km) should be avoided since birds that have been trained from long distances do not fly well from nearer home. Why this should be no one seems to know, but most experienced fanciers will not release trained birds nearer than 10–15 miles (16–24 km). One suggestion is that they fly over, going beyond their loft.

Obviously if the birds have had a particularly hard race or if there has been a long holdover then discretion must be used about sending them on a mid-week toss. Another advantage of a mid-week toss is that if a fancier can spare the time to watch them arriving or can get someone to watch for them, perhaps his wife whilst she is doing the housework, he will soon be able to pick out those that are consistently among the early ones. This will help considerably in picking out birds to pool.

As has been mentioned before, increased basket work particularly for young birds and yearlings is all helpful training, enabling them to settle down more easily and not to waste their energy worrying to and fro in the basket. One of the things most noticeable when attending the marking of the National Flying Club or the North Road Championship Club, apart from the superlative condition of the birds, is the way they settle in the baskets. No sooner have they been marked and put into the club panniers than they are settling down and nestling among the wood chips.

Lazy Birds

The last advantage of mid-week tosses concerns the problem of the lazy bird in the loft. Most fanciers will say that a bird that will not race and about which there is doubt should be got rid of, but there is usually in most lofts an old veteran, an old favourite, or one that is given just one more chance. These, if they are allowed out at exercise time, quite often fly as well as the rest but sometimes will be found to be pitching earlier than the race team and dragging the rest in. In many of the lofts where the fancier says his birds do not take enough exercise, quite often this is the reason. A careful watch will soon pick out the offenders and they should not be allowed to exercise with the others. The fancier must also harden his heart and get rid of them or provide them with a separate place for a quiet retirement if he has got the space.

Quite often the fancier makes it much worse for himself by driving all the birds out of the loft for exercise whereas the older ones which are the first to drop would, if given the chance, stay in the loft and therefore cause no harm. If the birds are basketed for a mid-week toss then the problem of an old or lazy bird dragging them in does not occur. They all have to fly the same distance and the old ones can be left at home and the lazy ones will soon be found out by being the last arrivals. As an argument against mid-week tosses it is said that this causes extra handling of the birds and that this is bad as it makes birds shy of trapping. This is the wrong approach to the problem; birds must be handled frequently and gently so that they are glad to trap and to be home. Nevertheless the cost of the mid-week tosses must always be considered, for if the birds will exercise well round home it will cost the fancier nothing in petrol.

FOUR

The Pigeon's Home

*Comfort – Fresh Air – Damp – Open-Door Trapping – Wildness –
Hand Feeding – Rough Handling – The Happy Loft – The Right
Size – Three Compartments – Trapping – Keeping the Floor Clear –
Brick and Tiles – Aviaries – Building a Simple Traditional Loft –
Sectional – The Front – Sides – The Roof – Foundations and Floors –
Assembly – Nestboxes – Perches*

It is vitally important that pigeons are kept in first-class health
and have a proper environment in which to rear their youngsters. Considering this, it is surprising that the design of the loft
is so frequently neglected. It is true that good pigeons have raced
to a bad loft and there are many, many winners in dilapidated
and ramshackle affairs. When these fanciers apologize for their
lofts the only advice to give is not to change them since they are
winning as it is. If they change their lofts this disturbance to the
birds might easily ruin their chances of success for several years.
To those who are not winning the best advice is to get rid of
unsightly lofts and get something that is a little less likely to
cause the neighbours to complain.

Every week fanciers used to write letters saying: 'I live on a
council estate and the council will not let me erect a pigeon loft.
Will you see if you can help me?' In a great many cases permission has been secured to allow a loft of the right design and construction to be erected, but in many cases it has taken a lot of
hard work to convince councillors that pigeon lofts are not
always old sheds patched up with scrap and painted all the
colours of the rainbow.

There is a custom, which fortunately is dying out, of painting a
loft in black and white or red and white stripes, with each stripe
about six inches wide. While the owners will admit that often it
does not look pretty from the kitchen window, they claim that it

helps the birds pick out their loft when they are flying over. There is no evidence to support this theory, and there are successful fanciers living in the country who have painted their lofts dark green so that they are inconspicuous against the grass and trees in which they are set. Their lofts are pleasing to the eye and success has come their way. My own birds used to fly in to the roof of a house, one house in a terrace of identical houses and they entered through what are windows of attics in all the other houses of the row. I never had a pigeon make a mistake. Even the very youngest of pigeons did not ever attempt to go near any neighbour's windows. The pigeon does not need any assistance in order to see its loft for it must not be forgotten that it has been shown that a pigeon's eyesight is very much more accurate than a human's.

Comfort

The design of the loft for some fanciers is no more than a matter of choosing one out of a catalogue. Most of the lofts on the market can be relied on; however, there are different designs and some are slightly better than others. If the fancier is going to make a loft himself in view of the expense of purchasing one, he should start with a good plan.

He should always remember that he, as well as the pigeons, has to go into a loft and, being 6 ft 2 in myself, I know full well the needless discomfort that can be caused by having a loft without sufficient headroom. A successful fancier must spend many hours in a loft during the course of the year and if he is taller than the loft, he is giving himself that much less encouragement to spend the requisite number of hours. The loft should be a place that is comfortable both for the fancier and his birds.

Fresh Air

The fancier's comfort cannot always be considered, for pigeons do not mind the cold at all and must have, for their health's sake, plenty of fresh air. Plenty of fresh air is often thought to mean the wind whistling through the loft, but this is a very different mat-

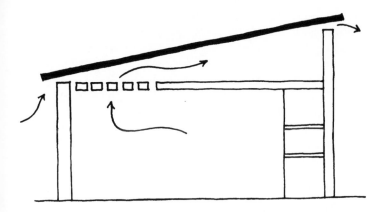

11 VENTILATION WITH SLOPING ROOF.

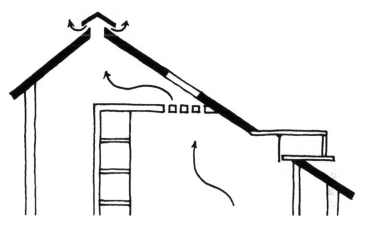

12 VENTILATION WITH APEX ROOF.

ter from having continuous draughts. To secure this circulation, the loft should have ventilation openings at both the highest and lowest points and on opposite sides. The front of the loft from about waist level will be dowelled and this allows the main volume of air to enter. This is the traditional Natural lot but not suited for Widowhood when during the racing season the front must be closed in.

It is usual on a small loft to have the roof sloping downwards from the back at a gentle angle and therefore ventilation holes should be drilled at the highest point at the back. As a rule these should be made so they can be open or covered according to the weather. At the same time, there should be openings at floor level. In most cases the gap under the door is sufficient. In this way, provided the loft is not overcrowded, the birds will get sufficient air and there will be a free circulation of air without bad draughts.

The actual design of the loft does not vary much whether the birds are to be raced on the Natural or Widowhood systems. The nestboxes may vary but the inside of the loft need not vary. The biggest difference will be in the accommodation for the hens during the Widowhood period and there are as many answers to this problem as there are Widowhood fanciers. Essentially the problem is that, except for race days, the cocks should not see or hear the hens. A separate loft in another part of the garden is ideal, especially if there can be an aviary, so that the hens can get some fresh air for they will not be flying out. Since the cocks should not see the hens when they go out for exercise, the aviary must be made in such a way that the hens can be cleared from the aviary and closed inside, while the cocks are out.

The box perches of any sort are out for the hens otherwise two of them may decide to pair together, so the choice is either V perches or, simplest of all, long poles running the width of the hens' section. This is the ideal situation but not every fancier can afford the luxury of a separate loft and the alternatives vary in their usefulness. A narrow loft can be built on the back of the main loft and can be remarkably successful. A loft section can be screened off by solid panels and with covered windows. I have seen this work remarkably well although in theory it doesn't seem such a good idea. One fancier even said that he kept the hens with the young birds and found that in spite of the temptations from young cocks the system worked. I would like to hear from others who have tried it this way. As we shall see when looking in more detail at the Widowhood system, in some of the variants the hens play very little part so this would make a lot of sense.

Damp

Damp and ventilation are continual problems and in recent times a wooden loft with a tiled roof has been favoured particularly in commercial lofts. The tiles are pantiles and are installed so that they are not a tight fit. This means that air inside the loft leaks gently out providing good ventilation without draughts. It is possible by using matching transparent tiles made of perspex or similar that roof lights can be included that are completely watertight. My own loft suffered from dampness but I modified it to include roof lights that caught the morning sun and this has greatly reduced the dampness problem. My roof lights are actually of corrugated transparent plastic because the loft is not tiled but the same would apply if it were.

In exposed positions there should also be some method of covering up parts of the front in winter or stormy weather so that the rain does not beat in and soak the floor of the loft and possibly the pigeons. Damp is the enemy of good health for not only does it mean the loft is much dirtier because droppings are spread, but it is also the breeding ground for many diseases. For this reason the roof of the loft must always be kept in good repair and where possible the loft should be placed so that storms cannot beat into it. In most parts of Great Britain the wettest winds come from the south-west and a loft should not face that direction, but rather south-east to east. In this way, of course the cold east winds can blow into it, but they will not do half as much damage as the damp. Whichever direction the loft faces, it will be necessary quite often to protect newly born youngsters from this cold east wind and, of course, the direction from which it is intended to race must also play a part in deciding the position of the loft.

If it is once decided that the loft should face south-east for climatic reasons, a fancier may have second thoughts in having his loft face away from the direction he is racing. If you are racing on the North Road, he may argue, it will save time if he is able to have his birds fly on to a loft directly. However, it is rarely that a bird comes out of the sky without so much as a turn or a half-turn and when it is dropping from a height it can easily turn itself to face whichever direction it likes without loss of time.

Open-Door Trapping

There are other ways in which the design of the loft can help in speedier trapping, and probably the most important of these is the system of the open door. On some lofts the birds enter either through drop holes or through bob wires, but as the name open door implies, in this system they fly right into the loft and to their nestboxes before they pitch. They do not waste any time by pitching on the roof of the loft or on the alighting board outside the trap. When they arrive on their nestboxes the fancier can be waiting to remove the rubber rings and clock them in. It is obvious that unless the birds are tame this system will not do, since the bird, if alarmed, can go out of the window or door as fast as it came in!

I have already spoken of the pros and cons of stall traps versus open-door trapping. A word of warning on stall traps. Not all work properly, the side walls must be made of metal or in some cases glass. Wood does not work because when the pigeon is in the stall, the wood will give its claws some purchase to get out again. Even pigeons that have entered the loft a hundred times with the bob wires allowing unimpeded entrance may well panic when the bobs are fixed. In an ideal situation the bird should wait patiently in the trap for the rubber to be removed, but birds returning from races are often highly strung and far from patient. Some fanciers sit below the trap itself on race days to save the vital seconds but even then control in the trap is essential. I don't like the idea of sitting under the trap because I like to watch the birds dropping out of the sky which may mean another second lost but then this doesn't really matter on the longer races. The stall trap has the obvious advantage for many fanciers in that it causes minimum disturbance in the loft. In spite of the fact that fanciers should be in and out of the loft all day in summer and whenever possible in winter, these visits should be quiet and peaceful. The danger is that when a bird comes home from a race particularly if the time seems good then there is the very human risk of rushing in to grab the pigeon. A stall trap will avoid this risk.

My problem is from prowlers. The loft is down the garden and the garden backs on to Hadley Common with huge old oaks and nowadays thick bushes that once were kept under control by the verderers. The trees are perfect for squirrels, crows and rooks and the undergrowth is perfect for foxes. You can keep squirrels out of the loft by closing off all large and small openings with small square mesh. My mesh is about $\frac{1}{2}$ in (12 mm) square and will keep out mice if they are a problem too. Squirrels will eat out of the hoppers and most annoyingly will just bite off the tip of the grain which is the source of the vitamin A that yellow maize contains. The spring, when young birds are being reared and the hoppers are most likely to be in use, is also the time when there are likely to be eggs in some nests and these are a great delicacy to squirrels. Spring is a bad time for crows too, and I am convinced that when they swoop low over the loft and 'buzz' it they are just trying to frighten the birds out of spite. Given the chance they will go into a loft through an open door or window and I have seen them tearing 3–4-day-old chicks to pieces. It is easy to say 'Don't leave doors open' but it can happen, for example, on a sunny day when the birds have just had a bath.

I did have a cat problem, a ginger and black tom that was always around the roof of the new loft. One autumn we had to go to the south coast for a pigeon club dinner and, as some birds were still flying, had to leave the trap open. When we got back there was a message on the answering machine saying there was a cat in the loft! Panic. But at the loft there were no bodies, just the birds at one end and a terrified cat at the other. The moment the door was open he was off and that has been the end of my cat problem. I am, however, investigating the possibility of installing an ultrasonic device. This is partly to combat cats but I hope it will also work with my big problem – foxes. I tried installing an automatic light that would flash on with any human or animal intruder, but this doesn't seem to work as I hoped. When the snow has been on the ground, I have seen the foxes' footprints as they have circled the loft at night, and during the day I have seen two different ones in the garden. I think this is the type of prowler that is most upsetting to the pigeons. Let's hope ultra-sonics work.

Some of the Widowhood fliers who do not use a stock loft will even insist that because Widowhood birds can race week after week, smaller racing teams are possible and just as likely to win. I am not entirely convinced by this argument if only because I know very few Widowhood specialists who don't have a good-sized stock loft. I think the statistics support me. Since Widowhood overtook Natural racing in this country, the number of fanciers has gone down and the number of rings sold has gone up. This seems to me proof that nowadays the average fancier keeps more pigeons. I don't know if my club is an average sort of a club but from what I see on marking nights there are not many small teams sent. There are fewer members that only send half a dozen or less but these were rarely winning out of turn.

Getting back to open-door trapping, some fanciers put a picket fence along the front of their loft so that the birds cannot pitch on the roof but must come straight in. Some birds will pitch over this fence but they are few. They can be stopped by heaping branches and gorse on to the roof.

There is some discussion on the way a roof should slope. Some say towards the back so that the rain is carried away from the front and so that the birds are discouraged from landing on the roof and pitch straight on to the trap. Other fanciers favour a roof sloping forwards so that when birds land on the roof they can be seen. By allowing sufficient overhang the water will not drip into the loft. A loft with an arched roof combines the disadvantages of both unless it is steeply sloped.

Some fanciers modify this system by having a single pair of bob wires over each nestbox so that when the bird enters its box it is kept there until the fancier releases it. This is obviously an advantage with birds that have a tendency to flutter round the loft, but of course valuable time can be lost in getting the bird from inside the nestbox. If the full open-door system is used the nestbox has the front removed when the bird is expected and the bird will pitch right inside the nestbox and can be quickly gathered up.

Wildness

The problem of wildness and tameness in pigeons is one which depends largely upon the amount of time a fancier can and will spend in his loft as well as the problems of cats and foxes. In a small loft where the fancier has no more than ten pairs of birds and where the birds are used to his presence, it should be possible for the owner and indeed for the owner accompanied by a stranger to walk into a loft without the birds moving off their perches. In many lofts this is very far from the case and on the approach of the owner the birds will flutter backwards and forwards and in a moment the whole loft is in a panic.

There are many causes of wildness in pigeons: too many pigeons for one man to cope with easily; too much space in each section so that they can fly round and round the fancier without ever really coming close to him. In a loft such as this there is not the close contact which should exist between a fancier and his birds. I have been into many, many successful lofts in the course of my work and have handled a good number of National winners and winners that have earned a place in pigeon-racing history. Without exception these birds were quiet in the loft, easy to handle and seemed to take pleasure in their owner's presence. I cannot think of one loft where the birds have been wild, difficult to catch or difficult to handle even if not all the birds in winners' lofts have been uniformly tame. Some strains of birds are more highly strung than others. Some lofts which are separated by some considerable distance from the fancier's house contain birds which are not quite so tame, but they have always been a very long way from being wild.

Hand Feeding

Wildness is not an easy thing to cure but it can be done by patience, and in time the birds will become used to seeing the fancier in and around their perches. It is here that hand feeding of the birds has its great advantage. Many fanciers, because of their work, are not able to give the amount of time they would like to attending to their pigeons and they must use the hopper system of feeding. The hopper, usually round with a conical top, is put in the loft and is left there either all day or for selected portions of the day, and the birds feed themselves from the hopper as they like.

In hand feeding the birds are fed in a tray or on the floor of the loft, provided the floor is clean enough. The food is given to them handful by handful, and the birds must eat it when placed before them or not at all. The birds can even be fed actually from the fancier's hand and made to pick out the corn from his palm. This has the great advantage of bringing the birds near to the fancier and, of course, is a great help in building up the bonds of familiarity. It is one of the most successful methods of taming wild pigeons, since they must approach a fancier in order to eat. A mixture is useful for hand feeding, since the birds often prefer certain grains of those in the mixture, and will hurry to him in order to eat them first.

Rough Handling

The biggest single cause of wildness is probably rough handling as youngsters or on returning from a race. In the shorter races, where the race may be won or lost on a split second and where it is not uncommon for £5 or £100 to turn on that same split second, the feelings of a bird are sometimes forgotten in the haste to get the rubber ring off and into the clock. This is a basic mistake and if continued will be the fancier's undoing. A bird has been trained to come home, has been encouraged to enter the loft with promises of food, its mate, or the pleasure of returning to its eggs; it enters the loft expectantly and is descended upon by a savage fury that wrenches at it and then casts it aside. Is it to be wondered that some pigeons give up wanting to race and if sent to the race point are just happy to jog along and arrive in their own time? The seconds saved in the clock are not worth a bird lost. Gentleness in catching the birds and gentleness in taking off the rubber ring must be the fancier's ideal.

Many fanciers recognize this difficulty and tackle the problem another way. When the birds arrive, they time in only their first one or two or their pool pigeon and never take the rubbers off the others, preferring not to handle them. This is one solution to the problem but this approach is from the wrong angle. The solution lies rather in handling the birds so gently every time they are caught that they will not mind being caught another time for the rubber to be removed. The constant handling of birds as often and by as many people as possible is really an advantage. Some Belgian fanciers will not allow anyone to touch their birds during the moult, but this cotton-wool technique is neither necessary nor desirable. It is, however, essential that every time the birds are handled they are handled gently and the comfort of the bird is considered every moment. The danger time is a hurried basketing for a training toss and I have on occasion basketed the birds by torchlight so they are quiet. It has the additional advantage of getting the birds used to spending time in the basket.

The Happy Loft

Handling may have an adverse effect on the plumage and in many cases may possibly accelerate the moult, particularly if the handler suffers from sweaty hands. But it should be obvious that no matter how good the feathers, no matter how perfectly the moult is proceeding and no matter how good the condition of the pigeon, the bird will not race successfully unless it has the will to return home and the will to do it at top speed.

There are many ways of giving a bird the incentive to race and these will be discussed in the chapter on racing methods, but unless that pigeon knows peace and happiness and is happy about his loft and his master, then these other methods will prove of little avail. There will come a day when that something extra is needed to get the bird home, a hard race, a long race, a difficult race; all of these will demand that something extra which only the fancier can give his bird.

Although this may seem a long way from considering the design of the loft it is in fact very relevant, for during the racing season the fancier should try to spend an hour or two most evenings with his birds. If he has to creep through narrow door-ways, crouch under a low roof or in any other way make himself uncomfortable, he will soon find that the eagerness to spend the extra ten minutes with the birds will go. It is that extra ten minutes that wins races.

The Right Size

He should therefore make his loft tall enough for him to stand anywhere in it with comfort. The height should be an inch or two over the fancier's own height, making allowance for a hat or cap if the fancier wears one. The width of the loft will be governed by the number of pigeons a fancier wants to keep but it should not be so large that pigeons can always be a couple of yards from him. The maximum size must be such that by standing in the centre of each compartment in the loft he can touch all four walls without moving his feet. In this way he will be able to get to his pigeons with ease and without scaring them unnecessarily. A floor size of 6 ft by 6 ft (2 m x 2 m) should be the maximum, but as this would not give enough space to keep a team of birds, several compartments are essential. It is a little misleading to say how many birds should be kept in a certain space since the number will vary with the ventilation and position of the loft, but in an average position with good ventilation a space of 36 sq ft (4 m²) is only suitable for twelve birds complete with their six nestboxes and perches fitted. Each compartment should there-fore be made this size and two or three compartments made in the loft. Three compartments for example will give a loft 18 ft by 6 ft (6 m x 2 m). This gives room for thirty-six old birds. In many races, highest honours have gone to fanciers who have kept even fewer birds than this. A large team is not essential for success. However, one can often hear of fanciers, the unsuccessful ones I should add, complaining about those who keep a large team: the mob fliers.

The animosity with which the so-called mob fliers are greeted is based on the misapprehension that they stand a better chance of winning prizes and pool money. In fact there are very few, if any, mob fliers who take more money out of the club than they put into it. If the argument was that mob flying is not true pigeon flying, there would be a lot more justice in it. It remains true that a few pairs of Natural birds or a small team of Widowers properly handled are cheaper to maintain and more likely to bring success than a large number of birds expensively maintained.

Three Compartments

A loft with three compartments is well suited to the needs of most novices. In winter the fancier will want to separate the cocks and hens to prevent breeding. During the racing season he will want to have his young birds separated and he may also, of course, have some pigeons which he keeps only for stock. With three compartments he should be able to house his birds comfortably and easily, to arrange them as he requires. There are extra problems if he is going to race Widowhood but I have already written about these. If the open-door system is used there will be windows or doors facing each section through which the birds will enter and leave. Some fanciers provide a special small section for trapping which opens into the others. A young or inexperienced fancier, in spite of the advantages of open-door racing, would be well advised to use some sort of trap until he and his birds have more confidence in each other, since in the excitement of a race the best intentions in the world are liable to be forgotten.

The small section cannot be ignored even with open-door trapping. I know there are some fanciers who say they wait at the loft until every bird is back. I don't believe them. You cannot do it for the long races and even in the short ones you can never be sur one won't catch a wire or get taken by a hawk so that you could be waiting for days. Of course it is a good idea to wait for most of them from a shorter race. You want to welcome them back and you want to see how they look, because if they look distressed you want to know the reason why. For the very late arrivals you either need a single drop hole in each section or to let them come into a centre section. This same centre section can contain the corn bins so it is not space lost.

Remember I am trying to describe the smallest and cheapest loft for a beginner. If you go to the British National (Old Comrades) Show you can see lofts there at prices up to £15,000 and they are splendid but unless you have taken early retirement with a huge golden handshake, these are not for my novice reader. I would not actually advise a beginner to buy one in any case. A fancier adapts to his loft and also adapts the loft to himself. For my loft, what I call the old one, I bought a commercial timber loft. Since then I have altered the floor, altered the roof, altered the traps three times, and fitted ventilation changes on all the walls. Each time there has been a small improvement but the cost of making the changes (except for the new roof) has been very small because the loft was sectional and made of timber.

For the benefit of the DIY experts I will describe the construction of a small sectional loft later. I should add that I do not think of myself as a DIY expert even though I have got tools, and even a Black and Decker Workmate. The plans are therefore based, like so much in this book, on the ideas and methods of those more expert. Before you hammer a nail in, try and visit some local lofts for what works in a sheltered valley in the soft south will not work on a windswept hill in Scotland. You will save yourself a lot of heartache if you modify my plans before building rather than changing them as soon as they are up.

Trapping

If only one trap is fitted to the loft, it should be fixed to the centre section and every bird should enter and leave through it. An old and still very good type of trap is the 'Piano' type. Its modern version, the 'Sputnik', closely resembles its predecessor of 100 years ago! Its advantages are that, first, it provides a platform on which food can be thrown to feed the birds. Second, the box part can be used by squeakers so that they can have some idea of the outside of the loft without actually going out. A third advantage is that if it is the version where bob wires are fitted there is no perching place near them which would allow a cunning bird to escape. The fourth advantage is not so great as some believe and

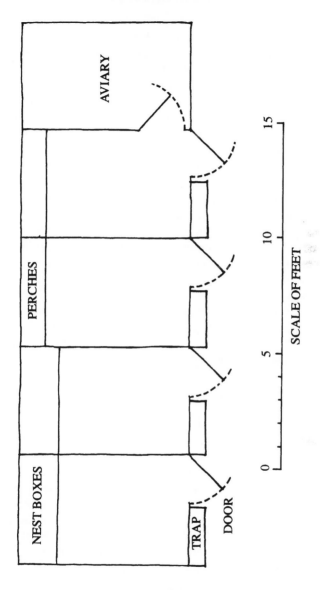

13 GROUND PLAN OF SIXTY-BIRD LOFT (BASED ON MY NEW LOFT SIMPLIFIED).

that is that doors can be fitted on the back so that returning birds can easily be caught. This benefit is limited, for the fancier should be able to catch birds easily and in any case they will soon learn what the shut doors mean. It is better to do without these doors unless it is vitally necessary. The even more modern so-called 'Yankee' trap has been described earlier. It is not uncommon to see bob wires fitted to the front of every section of a loft, but here trouble may arise when the bird arrives perhaps unnoticed and enters the wrong section. A young bird for example may be used to the right-hand section, and then the following year, has to be trained as an old bird to enter the left-hand section. This is putting unnecessary difficulties in the way of the pigeon and the fancier, and if possible the birds should enter the same trap as old birds and young birds. If the trap is in the centre then it is not difficult to arrange partitions so that as they go through the trap they can be diverted to their proper section. One arrangement is to have a small space behind the trap opening out to all three sections and separated from the outer section by sliding or hinged doors and dividing the middle section in two by more hinged doors. These, of course, like all internal partitions should be made of laths of about 1 in (2.5 cm) thickness and separated by 1½ in (3 cm) spaces through which the air can circulate.

Keeping the Floor Clear

In a small loft one thing that is likely to hamper the fancier as he moves about is a large amount of clutter on the floor. Pigeons must have food, water and grit, and many fanciers give them special foods and minerals in addition. If these are all kept on the floor of a small loft, there is not a great deal of room to move about. Nowadays the drinking fountain and food hoppers you can buy are nearly always plastic. The old Eltex enamel hoppers have gone the way of all old-fashioned long-lasting products. If you see any second-hand being advertised go for them. They usually only turn up when a fancier has to give up and they, like his baskets, are worth having. There are few lightweight galvanized hoppers about but they are only a little better than the stout plastic ones. The thin plastic bowls flex and crack when full of water but worst of all the cone-shaped covers of plastic fall off so easily. You can leave the drinker full of nice clean water and come back a few hours later and find the cover knocked off and the water filthy.

I am all for birds having frequent baths but not in the drinkers. It is amazing just how much water gets splashed around. A few years ago we threw away an old fridge but before I took it to the tip, I pulled out the plastic-covered wire shelves. These are marvellous under the water drinkers for letting a little air circulate and stopping the dampness settling under the drinkers. I have to use these more than I like because I prefer the method of outside feeding used very extensively in chicken houses, whereby small boxes with flaps on are fitted on to the outside of the loft and in these the food and the water are placed. The wall of the loft of course is cut and bars fitted so that the birds may put their heads through to reach the trays in which the food and water are kept.

In this way it is almost impossible for the birds to soil either the food or water, which may be readily changed without disturbing the birds. There is yet another advantage in this system. Although we have said that the successful fancier is one who spends a large amount of time in his loft, there are some days when, because of holidays, sickness or some other reason, he will not be able to attend to his birds. In these cases he must ask someone else to do it. If, as with this arrangement, all that is necessary is filling up the troughs from outside the loft, then it is not likely that a great deal can go wrong. The birds will not be disturbed by strange people walking about in the loft and there is no danger of the birds accidentally getting out or into the wrong section. This system of having external food and water supplies is used by many successful fanciers and is usually very popular with their wives.

Inside the loft, if traditional nestboxes are used they should have a much wider entrance than usual. This is to reduce the amount of jostling in the entrance to the nestboxes. In the event of the bird making a mistake by entering the wrong nestbox and being set upon by the rightful owner it is able to make a speedy

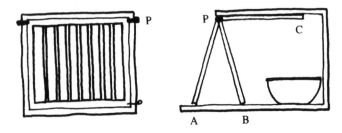

14 SWING-FRONT NESTBOXES.

getaway before a fight has developed. Fighting in a loft can be very serious and anything to prevent it should be welcomed. The old pattern of wooden nestbox front has a high entrance to prevent the young squeakers falling out. If the birds are removed as soon as they can feed themselves then there is little danger of this.

Inside this type of nestbox there is one nestbowl placed in the corner as usual, with sawdust sprinkled with a pyrethrum-based insecticide. When the birds are sitting their first round of eggs the practice is followed of putting up a couple of bricks or a piece of wood to act as a screen for the nestbowl and to give the bird a little peace and privacy. By this method it is possible to have the front out of the way except when it is needed. A single door of dowels is hinged at the roof. It can be fixed in three positions. The first (C) is up against the roof out of the way. This is the normal position. The second (B) allows a small perching space on the front. This is used at pairing up with the hen inside and the cock outside. It can also be used when the boxes are closed for winter. The third position (A) is right forward, preventing any bird from perching. This is used when one bird of the pair is away racing to prevent other birds worrying the bird left behind. This has the advantage of allowing smaller nestboxes.

Nowadays I am inclined more and more to the plastic Widowhood nest fronts even for Natural racing. The only thing I don't like about them are the floor fixtures, which are always difficult to clean round, but this is a minor point. The nestboxes will need four of these fixtures and the ones on the sides can be fitted to the wall not to the floor but this does need greater accuracy. The fronts come in several colours and I use mixed colours. One box will have a blue front, the next red, then yellow and green. Pigeons are said to be colour-blind. I don't know if that is completely true but, even if they are, then the colours will be recognized as different tones of black and white. In any case it does no harm, may do some good and costs nothing extra.

These fronts can be used in various ways: first, for pairing up. As described briefly before, the hen is put in one side separated from the cock by the partition with the sliding door. One pot of water and one pot of food by this central partition can be used by both. The loft section must all be paired up this way at the same time so there are no birds spare in the section. Depending on the time of year and the randiness of the cocks so the section will be full of loud cooing. Do not be in a hurry. In the case of a pair being put together for the second year, if the birds have been separated for a few weeks, the hen soon, possibly within minutes, will start to respond to the cock. Still don't be in a hurry but wait until you are sure and then let both birds out into the loft section. I don't just open the door but take off two sections of the front so they can get out and in easily. Watching them on the floor it is soon easy to see if they are going to make a match of it and treading may occur on the first day.

If I am pairing up more than one section of birds at a time then I can be doing the same in the next section. When I have watched the cock tread then both birds can go back into their nestbox. I put in a nestbowl behind the plastic front that is still in place and then if they do not fly up of their own accord they can be chivvied up. The food and water pots are put back in but this time in the corner farthest from the bowl. I add a third bowl which is a mixture of grit and the additive Hormoform. One fancier in Belgium uses exactly this system but with one variation. He has a sheet of cardboard (dark plastic would do) fixed to the wires on the side of the nestbox where the nestbowl is. He says he thinks that it encourages the hens to settle on eggs for this first round and he even uses it while racing Widowhood later. I think it is an idea worth trying.

Once one pair has been settled the next pair can go through the same procedure and with luck they will be easier to settle having watched the first pair tread. It can be a long job. If you start by putting them in the nestboxes Friday night you can be satisfied if they are all settled by Sunday night. Don't plan anything else for that weekend. The person who taught me so much about practical clock-setting, Jack Dixon, used to take a week of his annual holiday to pair up. His wife, very wisely, took a week in Spain! This makes sense because it is a worrying time and it can be stressful and time-consuming, particularly if one pair just will not go together. This awkward pair can be kept locked up if all the other pairs have got together so that you can confidently take off the front and allow all the settled birds out at the same time. It is still a good idea to feed them all in the pots, particularly if you are going to race Widowhood after rearing.

If you are going to race Natural, the fronts (even if called Widowhood) are left with just the one section guarding the nestbowl. The same is true if after this round has been bred the cocks are going to be put on Widowhood. Either way I prefer water and hard corn to be in the little pots in the nestbox but as soon as any chicks are hatched I add a hopper full of pellets and a drinker of water because babies in the nest need a lot of water – more than the pot may supply. I also add a pot of grit. I know some fanciers have grit and minerals in the loft all the time. I don't like the idea because I think it gets damp and unattractive. I put in half a pot every two or three days and if there is any left I throw it on the garden.

Bricks and Tiles

When a fancier comes to think about building his own loft then he may consider building materials other than wood. A brick-built loft is obviously a good sound job that can be made a really fine-looking structure. Many fanciers have raced successfully to brick-built lofts and providing they are properly constructed and have adequate ventilation there should be no danger of dampness. Concrete lofts are not popular partly because of the difficulties of construction but there are available now prefabricated concrete blocks which can be used to build up a loft. These fit tightly without having to be mortared. The trouble with concrete blocks and to a lesser extent with bricks is that they are cold to the touch and tend to encourage damp. For that reason they are not really suitable for a floor and if it is necessary to use a building which has an existing concrete floor then it should be boarded over. Good ventilation, of course, will reduce the danger from dampness.

Tiles have increased in popularity in recent years. Again I think it is the Belgian influence where there are still lofts in house roofs and many more brick-built permanent (and expensive) lofts. If you use tiles in the loft, do not fit them as you would to a house. Dutch-style tiles are used and are unlined, so that with a loose fit there is a slow ventilation effect through the roof. They are also good for being cool in summer and warm in winter but they are a heavyweight roof not suitable for a lightweight wooden loft.

Asbestos sheeting is a useful material, not being quite so cold as slates, tiles or corrugated iron. It has the disadvantage of being breakable if something is dropped on it and must be handled carefully while being fitted. It can be used successfully instead of roofing felt over wooden planking. Corrugated iron was much easier to fit but it was noisy in bad weather and the most unsightly of materials. Properly used and properly maintained it can make a good roof, providing, as in the case of asbestos sheeting, it is built on supports. I am really talking in the past tense here because corrugated iron is not as it used to be. It can be obtained only with difficulty and is then more expensive than asbestos or plastic. If you find a job lot of second-hand corrugated iron it could be well worth using. Corrugated plastic is probably the material of the future. It is lightweight so can be used on a wooden frame loft and because it comes translucent it can be used to add light and heat to the loft.

My older lofts face east and therefore get the morning sun. When I had the roof altered I had a steep sloping corrugated plastic front and a longer, more gentle, slope towards the back.

This was felted but it could have been opaque plastic. It has worked a treat, making the loft warmer, dryer and better in the winter. It has also had one effect not anticipated. I have always had lights in the loft, mostly so that I can feed and water in the winter months. With a time switch they can be used for early breeding but I have always had the feeling that one day I might want to have lights on for a late-night arrival from the longest race.

Ever since I read about the American Army lofts that had races in Hawaii at night I felt encouraged to experiment. Many fanciers like myself gave the birds exercise flights and even training tosses late in the evening so that the birds would be encouraged to keep going at the longest race. A good many fanciers found this worked but I wanted to experiment and go farther. I exercised the birds later and later and they would keep flying until I put the loft lights on. It was not pitch black because the loft was on a main highway with powerful street lighting. I even gave them training tosses but I only got to a few miles before I gave it up because the birds were skimming the roof tops and some were injured on TV aerials and, second, I decided it was unsettling them. As they swung round overhead you could hear the wingbeats and they were very short and staccato. The birds did not seem particularly nervous but these wingbeats did not seem natural so I abandoned night flying.

However, I did not abandon my idea of having lights available for the longest races and my plastic roof with the lights inside shines like a beacon in the summer dusk. One day I am sure it will give me the only bird on the day at 500 miles! Well, at least it means that when I go down the garden in winter I don't trip over anything in the garden. By way of final explanation I should add that the masterswitch and now a circuit breaker are just inside the kitchen door so that I can switch them on before I leave the house. I think that explains why I like transparent plastic corrugated sheets for at least part of the roof. One final point: the easiest to obtain sheets are almost invisible to birds, although they soon dirty up, so I have always used them with a ceiling. Since the roof in any loft will slope, the inside should be covered by a level lath ceiling. Wire netting can be used instead of laths but then it is essential to make the mesh small.

I have seen wire netting used on aviaries, with mesh roughly the size of chicken wire. It was disastrous because a bird could push its head out but then not get it back because the feathers would not give. Usually birds worked themselves clear in time but not until they had had a disagreeable and frightening experience. Any wire netting or any similar mesh used should never be more than $\frac{1}{2}$ in (12 mm) size. At this size it will keep the sparrows out too. The inner lath ceiling also prevents the birds getting up out of the fancier's reach at the higher points of a loft. Since it is non-continuous it allows air to circulate freely. If the ventilation holes are placed above the level of this ceiling then the birds are kept away from any drafts. When a corrugated roof is used on a wooden loft it is possible to save the expense of a wooden under-roof and just use this lath ceiling.

Aviaries

Aviaries, flights or fly-pens, are all too often ignored by fanciers only keeping a small number of birds but even in these small lofts they can be most useful. In the larger lofts they are almost essential. The aviary consists very simply of a 2 in x 2 in (4 cm x 4 cm) wooden framework with small-gauge wire netting fixed to it. It is built at the side of the loft and usually joins it by a small door. Sometimes the aviary will be covered so that birds can use it in bad weather, but normally it will just have a wire-netting roof. It need not be very big and the size will depend on the uses intended for it.

What are the uses? First, it is an exercising pen for those birds which cannot be liberated since they have been bred and raced from another loft and are being kept for stock. These can be broken into their new loft, but normally only when they have been paired up in the spring. The second use of the aviary is as a sun room. Instead of giving birds an open loft so that they learn to sit about the roof tops, the birds can be let into the aviary and can rest there soaking up the sunshine. It is surprising that when there are such advantages to be gained from a small aviary more are not in use. The use of an aviary will also be mentioned in the section on Widowhood.

One difficulty is that the ground on which an aviary is built can become muddy and unpleasant if care is not taken. There are several solutions to this problem: one is to dig the ground and plant green vegetables for the birds to eat; another is to cover the area with gravel; and a third is to concrete the space over, a permissible use for concrete. Inside the aviary there should be plenty of perches for the birds so that they can sit undisturbed by other birds flying round trying to find somewhere to settle. Food and water should be kept inside the loft itself since it may not be possible to use the aviary every day. Some Widowhood fanciers use an aviary with a roof for their Widowhood hens. The hens are shut in a small loft while the cocks are out exercising but for the rest of the time the hens are shut in the aviary.

The ideal perches are perching bars. These are rods and I have seen them made out of old broomsticks. They are better than V perches which can be fitted only on the walls. Box perches are not suitable because the hens may turn lesbian. A mesh floor is an advantage because it discourages nesting by two hens on the floor. I shall be talking about the handling of Widowhood hens later, but suffice to say that the more fresh air they receive and the fewer potential nesting places they have access to the better.

Building a Simple Traditional Loft

A beginner just starting up in the sport will probably not be able to afford a large loft and it is to tide the novice over his first few years in the sport that the following plans for a 6-ft (2-m) loft are given. It is planned for Natural racing for I am firmly of the opinion that the novice should start by racing Natural even if he turns to Widowhood after a year or so. The reasons are fairly obvious because success in Widowhood relies on interrupting the Natural cycle so the novice must understand what it is that he is interrupting.

The size may seem unreasonably small but it is deliberately so because the fancier must beware of overdoing it in the first years. Once he has mastered a 6-ft (2-m) or 12-ft (4-m) loft then he can consider expansion. This loft will not include a centre section but it is built in sections so that the fancier working in his spare time can build one section at a time. It is not a grand affair and the final quality will depend a lot on the skill of the fancier with hammer and saw. All the builders we have in the sport will probably not think much of the methods, but they are only suggested for a temporary structure with perhaps a few years' life. More particularly, the loft is something that a man without training can make providing he can saw a straight line and put a screw in properly.

Since cost is an important consideration, the design given is for a lean-to, to avoid the cost of the rear wall. As the loft is built in sections, a rear wall can be put in when time permits. In some cases a lean-to should be avoided. A lot depends on what sort of wall it is going to be built on. For a start, a loft should not be built on to a wall of a house since not only can the food attract mice, but also the fancier's wife or mother will be very rapidly turned against pigeons after the first moult. Even if the other occupants of the house can be pacified, it is unlikely that the local council and public health authorities would wear it. It varies from council to council but as a rule of thumb nowadays they like them to be at least 30 ft (10 m) away from any house. Nowadays gardens don't have brick walls except in small town gardens but the loft could back on to a garage. If the loft is fixed to the wall of an out-building then the roof of this out-building should slope away from the side where the loft is. If it slopes towards the loft then even if there is a gutter that works, the rain can still fall off this on to the fancier's loft causing further dampness and in time putting undue wear on the roof. A garden wall can be used if it is high enough and if it is not high enough then the top section of the rear of the loft should be built up on the existing supports of the sides. Trouble may be found in making the rear wall joint watertight.

Sectional

Even with this size of loft it is important to make the building sectional. The first and most important advantage is that you may not be charged by the council for it. If a fancier has a brick building or what is decided to be a permanent building he may expect to pay a charge to the council on it. If it is sectional and portable then he may not. Another advantage is the ease with which it is possible to extend the building. The plans given are for a 6-ft (2-m) loft but when a fancier needs more room he can join a second 6-ft (2-m) section on to the first and be sure of a good match. With 12 ft (4 m) of loft this will probably be enough for him for some years. Some of the sections are sufficiently small to be made indoors during the winter months, providing you can secure the co-operation of the other people who have to live in the house.

The Front

The construction of the front is not difficult. With a little trouble a fair attempt at a halved joint can be made. Two 2 in x 2 in (50 mm x 50 mm) posts should be used and if a piece 2 in x 1 in (50 mm x 25 mm) is cut out of the end of each post the two posts can be joined together with a tightness depending on the accuracy of cutting. They are then nailed through the joint. If a mistake is made then the joints can be strengthened by angle-irons, although this will not improve the appearance of the job. The cross-piece running half way up along the front of the loft should be made with well-weathered wood. This wood must not warp for it is in an exposed position. The top half of the front will contain the trap and the windows. A simple alighting board can be made to fit on the half of the frame farthest from the door. The nearer half can be fitted with dowels either fixed or fitted into a frame on hinges that can be opened. The alighting board is fitted to the corner upright and should be fixed after the loft has been completed.

The bottom of the front is covered by boards fitted overlapping, starting from the bottom. Here again tongued and grooved boarding is best but more expensive. The use of cheaper weather boarding will reduce the cost of the loft quite considerably but the danger of this boarding is that when youngsters are on the floor they may be subject to draughts since even the best weather

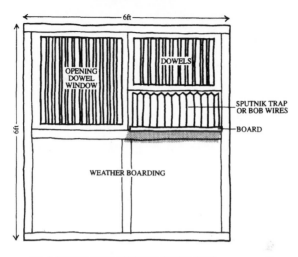

15 A SIMPLE LOFT SECTION: FRONT VIEW.

boarding cannot give a perfect fit. It is usually easier to deal with these draughts by internal pieces of plywood or plasterboard fixed on the inside of the loft. The planks extend 2 in (50 mm) either end beyond the framing on the front only so that when the loft is assembled this overlap covers the sides of the uprights of the next section. When the front has been completed then it should be given an undercoat of paint and when that is dry, a second coat. The whole of the front should be painted, including those parts which are going to be concealed by joints later.

Sides

When the front has been completed then work can be started on the sides. On one side a door will be fitted, the other being completely boarded. The only places that are likely to give any trouble are the joints between the top frame and the uprights. Since they do not meet at right-angles they must be made as halved joints at an angle and nailed. If they do not seem secure enough then it is possible to screw a triangle plate on the inside of the framework. The boards are nailed as on the front and the whole painted. On both sides as well as on the front, it should be noted that the bottom plank of the weather boarding overlaps the bottom edge of the floor and extends for that reason 1 in (25 mm) below the framework. Care must be taken in storing and handling the various sections since this bottom edge can easily be split off.

The side with the door is made next. On this side the centre upright is moved towards the front of the loft, so that a full-sized door can be put in. It is best to hinge the door on the support nearest the wall, and in a loft this size to make it open outwards. A door can be purchased in one piece or quite simply made by nailing ¾-in (18-mm) tongued and grooved boards on to ledges also of ¾-in (18-mm) wood, 6 in (15 cm) wide.

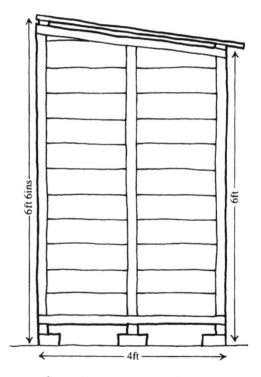

16 A SIMPLE LOFT SECTION: SIDE VIEW.

The Roof

To make the roof, three 2 in x 2 in (50 mm x 50 mm) battens run the length of the loft and tongued and grooved boards the same size as those used in the door are nailed to the rafters. Care must be taken that the nails are evenly driven home and the heads are flush with the wood. It is almost impossible with this simple type of building to be able to bolt the roof on, so it should be supported by wooden cleats sticking up from the top frame of the sides. The roof rests on these and is nailed firmly into place.

Although there are many possible ways of covering the boards, roofing felt is the cheapest and has much to recommend it. The roof itself is covered in the normal way but extra care is taken to see that the felt extends right over the edges of the roof and is nailed on the underside of the boards. It is a mistake to try to skimp with roofing felt since a leaking loft will soon cause the timbers to rot. The overhang of the roof should always be considerable since not only does it give greater protection in bad weather but it also improves the appearance of the loft. As a general rule there should be at least 1 ft 6 in (45 cm) overhang at the front of the loft and 1 ft (30 cm) either side. There is one exception to this, and that is when it is intended to erect an identical sectional loft very soon after the first. In that case the overhang can be cut down to 6 in (15 cm) either side so that when the new section is built the roof can be moved bodily to give a 1 ft (30 cm) overhang on one outer edge. The joint between the two sections must be covered by a 1 ft (30 cm)-wide strip of roofing felt like the top sealing strip.

If it is possible to make it a lean-to loft then this will save a little, but a free-standing loft is always better. A lean-to does have special problems – it is most important to make sure that there is no leak between the wall of the original building and the top of the roof. The best way of doing this is by laying a 6–9-in (15–25-cm) strip of felt along the top edge of the roof, raking out the mortar on the first course of bricks above the roof level. The roofing felt is pushed into this slit and the mortar made good. The roofing felt used for the rear seal is allowed to extend at least 6 in

(15 cm) down the roof over the felt used in covering the roof. In this way a good join can be made. When the felt is nailed down the nails should be put in at the lower side of the strip and the felt should not be stretched tight before nailing. In this way if there is any slight movement in the loft due to opening and closing of the door, this weather-proofing strip will not tear away.

Foundations and Floors

Before erecting the sections the foundations must first of all be laid. These should be brick pillars, a brick and a half square raised at least 12 in (30 cm) – if possible 18 in (45 cm) – off the ground. If you have the chance and the neighbours won't complain it is worth considering as much as 3 ft (1 m) off the ground. It will mean steps and perhaps a platform but it does give a wonderful amount of storage space. In this way a space will be left under the loft sufficiently high so that cats, mice, rats or other vermin cannot hide underneath without being seen. At the same time it will enable the air to circulate round the loft and prevent the damp coming up. About nine pillars are needed: one in each corner, one in the middle of each side and one under the centre of the floor. On top of these should be laid the main joists of at least 4 in x 2 in (100 mm x 50 mm) timber. The supports for the floor go on all four sides of the loft and this has the advantage of giving greater strength to the building, although it does increase the cost by increasing the amount of 4 in x 2 in (100 mm x 50 mm) used.

The standard practice is to have three joists running the length of the loft and to nail planks cross-ways on this. Since these planks run cross-ways the scraper will always be pulling across the grain of the planks and unless a top-quality hardwood is used this will cause splintering and will make grooves between the planks in which dirt will collect. If the planks are nailed on long-ways then the fancier is always scraping with the grain of the boards and will not tear up his floor so easily. A lot depends on how the fancier cleans his birds out; some are able to clean them out every day and if this is done then a small hand scraper can be used throughout the loft. Others can only clean out once or twice a week and then while a hand scraper can be used on the perches a long-handled scraper is best on the floor.

When the joists are down, the planks should be laid on. These should be 1 in (25 mm)-thick boards about 6 in (15 cm) wide. They should be fitted flush up against each other. If the fancier can afford it, tongued and grooved boards are better but an adequate floor can be made from those cut flush. Second-hand timber, if sound, can be used and scrap yards can often supply the right materials at a very reasonable cost.

When the framework of the joists is completed then it should be given two coats of creosote. Before floorboards are fitted the underside of the boards should be creosoted; the top is left in the natural state. The creosoting of the underside should be done with particular care since it may be many years before the fancier can get under his loft to re-do a bad job. Creosote is a good old-fashioned preservative. More modern ones can be used but remember most of this wood is going to be out of sight.

Assembly

All is ready to begin assembly except that for a lean-to rag bolts must be cemented into the wall to hold the uprights of the sides; two for each upright is sufficient. It is usually advisable not to drill the holes to receive these bolts until the loft has been assembled, since any slight inaccuracies that might have occurred in the building can be corrected. The holes have already been drilled in the bottom framing of the sides and these holes are continued through into the joists underneath so that bolts can be pushed through.

When both sides are firmly bolted to the wall and floor then the front can be added. The front boards will overlap the framework by 2 in (50 mm) either side, so that when the front is added the two uprights of 2 in x 2 in (50 mm x 50 mm) will lie one inside the other and will be bolted through from the side. The walls are now complete and the roof is added. Great care must be taken to see that this is done neatly and that the roofing felt is not damaged. This can best be accomplished with the help of other fanciers. The space between the battens on the roof must not be covered since this is intended to provide ventilation.

When the main structure has been completed then the door can be added and the dowel frames fixed. These are best made separately and added to the finished structure. Wooden dowels can often give trouble before getting them all fixed. If the window is made on its own to screw into the main frame then the assembly is easier. The top edge of the windowframe is drilled right through so that the dowels can be inserted. They are fixed by glueing afterwards. I have got bolts through my window sides with wing nuts so that I can fit plastic sheets over them for the winter or for Widowhood.

All the parts of the loft should have been painted beforehand and then all that will be required when the loft is complete is to touch up any woodwork that has been chipped. Once the external loft has been completed, then attention must be given to the internal fittings. There are a wide variety of hoppers on the market for corn, grain and water, and there is so little to choose between them that the fancier could purchase any without being in any danger of making a mistake. The only mistake he can make is to buy ordinary tins without covers, which would allow the birds to soil the food, grit and water, thereby opening the door to disease.

Nestboxes

One-half of any newly built loft must be equipped with nestboxes and here again there are any number of designs to choose from. A novice can purchase them in several different designs, usually in removable blocks of three and six, or they can be made to any design he likes. The advantage of removable nestboxes is that at the end of the breeding season they can be taken out of the loft and the whole loft can be thoroughly cleaned. Unfortunately they will be more expensive, even if the fancier builds them himself, than those which use the walls of the loft as an integral part.

The size of the nestboxes need not conform to any strict rules, but they must be large enough for the birds' comfort and for the

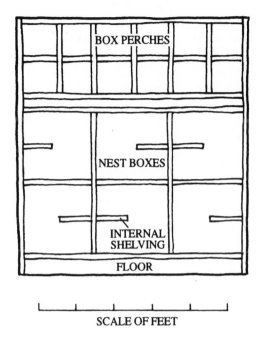

SCALE OF FEET

17 INSIDE VIEW OF THE BACK OF ONE LOFT SECTION.

convenience of both bird and fancier. An average size in a Natural loft is about 18 in (45 cm) wide, 15 in (38 cm) high and 24 in (60 cm) deep. The width of 18 in (45 cm) will just take two small nestbowls side by side, which is useful when the second round is laid. The depth of 24 in (60 cm) means that there is some space in front of the nestbowls where the bird not covering the eggs may perch. If space is left like this I do not believe it necessary or desirable to have any other perches in the nestbox section. A common practice is to have the nestbox slightly taller than normal, and halfway up along the back to fix a shelf. This has the advantage of providing a refuge for the hen if the cock is driving her hard and is also a suitable shelf where the nestbowl with the second round of eggs can be put.

With regard to the fronts of the nestboxes, the possibilities of

individual design are enormous. One type having much to recommend it has been discussed earlier. The more usual type has a
door made of dowels or laths and a small pop-hole in the middle. This pop-hole is covered by a door, hinged on the lower
side, so that when the door is opened the door itself provides a
perch for birds entering the nestbox. The advantages of the other
type of swinging nest fronts were also discussed earlier, but it is
worth commenting here that the old type with the pop-hole does
give less light in the nestbox, and although the idea must not be
carried too far, it is better for the nestbox to be a little on the dark
side than too light, particularly for those birds that are rather
nervous.

In the old-fashioned dovecots of centuries ago, it was believed
that if good pigeons were to be reared they should be bred in
almost complete darkness. This probably harmed the health of
the birds in other ways so no one would suggest complete darkness today. Nevertheless the quiet and restfulness of a shady
nestbox is undoubtedly appreciated by a breeding pair. The nest
fronts can be lifted out for cleaning. It is generally considered
inadvisable to disturb newly hatched youngsters in the nest
while they are being fed on soft food, and therefore for the first
week after hatching, the nestboxes should not be touched. After
that, however, the nestboxes should be cleaned out at intervals
since the youngsters themselves will be making a considerable
mess and if this is allowed to collect, particularly in hot weather,
it will soon be alive with insects. With removable fronts it is a
small matter to clean each nestbox out and nowadays you can
buy quite cheaply plastic corner shields to go behind the bowls
which makes cleaning easier still. On the other hand it is worth
putting the cart before the horse and making the nestboxes so
that the standard plastic Widowhood fronts fit them. There are a
wide range of these but the Boddy and Ridewood catalogue
advertised in *The Racing Pigeon* gives illustrations of most or all.

I have been saying from the very start that you should read
this book from cover to cover before spending a penny or doing a
thing. If I have repeated myself occasionally it is because every
step you take governs the next step and is governed by the previ-

ous step. I have suggested a 6-ft (2-m) section to the loft. That is 6 ft (2 m) internal measurement. I have also suggested 18-in (50-cm) nestboxes, that is external measurement, so that the section will take a row of four nestboxes. This is a reasonable size for Natural racing and a stock loft. It is not large and others might think 24-in (60-cm) (external) nestboxes more practical. In either case you would have to make the nestbox fronts yourself. The Boddy and Ridewood catalogue does list wooden nest fronts of the traditional sort but they are 19-in (48-cm) front width. The box would therefore be about 20½ in (52 cm). In other words, this ready-made front would not fit either of the suggested sizes. The current price at time of writing is £7.65 each compared with between £1 and £2 for plastic fronts, which makes them an expensive luxury. There are other manufacturers who may have different sizes and prices but the example gives a rough comparison.

Perhaps it now becomes clear why I suggested starting as a novice flying Naturally but using Widowhood boxes, even though these are usually much larger. The fronts advertised in the catalogue are 25 in, 29 in or 33 in (63.5 cm, 73.5 cm or 84 cm). The cheap plastic ones are in the larger sizes, and to me (as I have already confessed I am not a DIY expert), there is no contest. Two 33-in (84-cm) fronts with the boxes will be about 6 ft (2 m) and therefore would fit in with my plans but it does mean that there will be only half as many nestboxes in the section. The fronts are 33 in x 13 in (84 cm x 33 cm) so with top and bottom of the nestbox that's about 15 in (38 cm). Four rows would make 60 in, i.e. 5 ft (1.52 m). I don't like nestboxes on the floor, if only because it's too much bending and stooping, so I would lift the boxes off the floor and use the space at the bottom for storage.

This way you get eight pairs for racing Natural, and eventually eight Widowhood cocks per section. I honestly think that is enough for a novice, whether young or old, just starting in the sport. The other alternative is to make the loft with larger sections. There is nothing holy about the figure of 6 ft (2 m). If you wanted three 29-in (73.5-cm) boxes (internal) to a row you could make the section 7 ft 9 in (2.36 m) (internal) and even perhaps settle for three rows, less stooping and more storage space. The stooping is not just a matter of creaky bones but the difficulty of seeing what the birds are doing on the nest, etc. If you have boxes on the floor you have to crawl around on your hands and knees and the only advantage of that is that the floors are kept spotlessly clean. If you have time and DIY skills then it is possible to consider making your own fronts to whatever size you like but even then I reckon that the cost in materials is more than the £1.55 fronts I am talking about.

The construction of the nestboxes is not difficult if they are made as a block. First of all the space into which the nestboxes are going to fit is measured. This is divided by 18 in (45 cm) or 33 in (84 cm) or whatever, to see how many nestboxes it will be possible to accommodate. As these are being made to measure, the width of each nestbox can be adjusted so as to get an exact number in. Allowance must be made for the thickness of the wood used. The first job is to make from matchboarding or chipboard the side walls for the whole unit. If one wall is going to be the wall of the loft, then it will not be necessary to build another to go over the loft wall. When these walls are complete then a length of 1 in x ½ in (25 mm x 12 mm) fitted on these walls provides a support on which the floor can be built. Each floor in this way will be individually constructed from boards between side walls which run through from top to bottom. They should be strengthened at top and bottom by stouter wood so that they form a substantial wall.

When the floors and sides have been made, then it only remains to make each of the fronts individually and fit them in if that is what has been decided. It is not a difficult job, but time-consuming, and will be spoilt if rushed. The fancier will be well advised to draw up a rough sketch of his proposed nestboxes before he proceeds to make them, in order that he does not make any costly mistakes. It is always useful to have nestboxes spare over the number of pairs, since even if the number of birds is not increased, it is quite often helpful to have a spare nestbox in which to put strays or injured birds, or even for those pigeons which prefer to rear their second nest in a different nestbox.

Perches

While the first section of the loft is fitted only with nestboxes, the second section, used by young birds in summer and hens in winter, contains only perches. Perches come in many different shapes and sizes. There is the inverted 'V' pattern which has not been in favour for many years, largely because it is claimed that it deforms the keel of young birds. I do not think this charge is justified and this type of perch could well be used, well spaced out, on the solid wall of an aviary. Another type of perch particularly favoured by fancy pigeon keepers is a metal arm terminating in a wooden disc on which the pigeon stands. These are extremely useful in the hens' section but neither of these offer the advantages of box perches.

Box perches can be fitted compactly and, providing they are not built too small, prevent one bird from soiling its neighbour below. These can be very easily made by cutting boards with slots halfway across so that the uprights and side pieces inter-

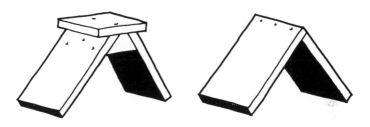

18 PLATFORM PERCHES (LEFT); SIMPLE V-PERCHES (RIGHT).

lock. These are strengthened at the corners and then simply put on the wall. A lot of these box perches can be obtained as prefabricated kits of chipboard and although I suspect a DIY using real wood could produce a more substantial job, the price would be that much higher even if the perches lasted longer. Quite a number of perches will be needed, at least twice as many as nestboxes, but since they will rarely be more than 1 ft² (30 cm²) each, the space occupied by twenty of them will only be about 4 ft x 5 ft (1.2 m x 1.5 m). I have one section of perches made up with backs. It is on keyhole plates and I hang it over the nestboxes in the winter. These are about 6 in (15 cm) deep. In the YB section they are much deeper – over 12 in (30 cm) for reasons that will be explained in the section on YB racing.

FIVE

Breeding Winners

Best to Best – Young Bird Tests – 'Well Bred' – Stock Birds – A
System Needed – Inbreeding and Crossbreeding – Which to Use? –
Magnification – Selecting Birds – Strains – Pure Nonsense – Breaking
to a New Loft – The Science of Breeding – Chromosomes and Genes –
Allelomorphs – Dominant and Recessive – Sperms and Ova –
Homozygous, Heterozygous, Genotype and Phenotype – Sex
Determination – Sex Linkage – Back Matings – Increasing
Homozygosity – Additive Genes – A Breeding Programme – Choosing
a Cross – Controlled Heterosis

The central problem of breeding is which cock to mate to which
hen. Many fanciers sidestep the problem completely and let their
birds mate up 'naturally' as they call it. They let the whole lot
loose together and allow them to pair themselves up as the birds
wish. 'Love matches' is a phrase frequently heard. The advantages of this system are, first, that it takes no time or trouble on
the fancier's part, and, second, that as the birds have picked their
own mates there is little trouble in getting them to settle down
and less danger of fighting breaking out. The disadvantages are
that if you breed a good bird it is only by luck and you are far
more likely to make the quality of the birds deteriorate.
However, remember what was said earlier about the difficulty, if
not impossibility, of racing and breeding at the same time. The
fancier must never lose sight of the fact that a happy bird is likely
to be a successful bird, and should be strongly in favour of 'love
matches'.

In fairy stories the two doves are always faithful to each other,
until parted by death. We fanciers have a less romantic view and
know that most cocks will chase any unprotected hen in the loft
so that we can create love matches. Using the Widowhood pens
in the way previously described these weddings can be arranged.

They work 9o per cent of the time and usually the unsatisfactory pairings can be switched to make them satisfactory. In the racing loft the emphasis must be on happiness but with the stock loft a little more patience may be needed to bring about the desired mating. Even then a fancier may have to resign himself to the fact that not every pairing will go as planned on paper. If the birds are yearlings it is worth considering the possibility that you have misjudged the sex of one of the birds. It can happen in the best of lofts.

Best to Best

What is obviously preferable is to select birds carefully so that the quality of the entire loft is raised and at the same time make it possible to repeat successful matings where it is wished. The obvious answer to the problem would seem to be to mate the best racing cock to the best racing hen and probably the majority of fanciers in this country do this now. It is the way which would seem to give the greatest hope for first-class youngsters. But there is a great difficulty about this simple system. If your two best racing birds are mated together then at some stage in the racing season they will both be feeding youngsters. If they are mated early enough to clear the youngsters before the first race then the moult will be too far advanced unless you separate them again. When they are undergoing this strain, you may not want to send them to the races. It means, therefore, that for that one or two weeks' racing your two best birds cannot be sent. Another problem it raises is, what are you going to do if one of the birds gets lost before the final races? Although they are your two best birds one of them may come to grief at an early race: a storm, a telephone wire, a hawk, a so-called sportsman with a gun may lay the bird low. Is the other bird to be rested for the remainder of the season? Is it to be raced unmated? Or is it to be found a fresh mate, not always an easy process?

From this it would seem that the birds to mate together would be one of first-class racing ability and one which is not necessarily a good racer but which is a good breeder. Since this may not be known, the bird should be well bred; that is, it should come from parents of known value. A mating based on these lines would be, for example, an old and experienced hen that it is intended to race 500 miles (800 km) on the Natural system that year and which will be specially prepared for that race, and a young Widowhood cock, a yearling or a 2-year-old, that will not probably have been raced beyond 350 miles (560 km). More probably it will be a yearling and not much will be known about this yearling cock for it can only be judged on its young bird performance.

Young Bird Tests

Young bird racing is not reliable as a guide to a bird's future ability. A young bird race in one local club was won by a fancier who had given his birds one toss in training. This youngster was carried with the mob right over his own loft, dropped out of the crowd, trapped quickly, was timed in and won the race. Without belittling this performance, how can it be a guide to what a pigeon will do flying ten times that distance, alone, and through rain and strong winds? From this it is obvious that young bird races are no sure guide to later performances and many a bird that shone as a youngster never shines again.

This is the crucial problem of the Widowhood system and one of the reasons that more and more interest is being shown in the Roundabout system. Some fanciers think of it as no more than a variant of Widowhood but in its treatment of hens it is completely different. Although I shall be returning to the subject, I want to look at it now in this particular context. If you keep a large stock loft this is not the biggest problem you have. Stock lofts will breed birds for the YB racing team in the first months of the year but they have to be fed for another 9–10 months. It is a very expensive luxury for a novice as his whole YB team has to come from the stock loft. As I shall explain later this does have its advantages. If you keep proper records then you soon know which stock birds are breeding the goods and which are breeding wasters.

Providing you are resolute and get rid of those that breed the wasters you will be on the right lines. This is what is meant by progeny testing but it only works if complete and accurate records down to every unhatched egg are kept. This is the problem that arises from buying expensive birds from the big breeding studs. Their detailed breeding records are not available to you.

Unfortunately, the novice, young or old, has neither the money, the space, nor the experience to have a large stock loft. He may have a few birds flying out or even prisoners that have done well for others and are now too old to race. If they breed well he should keep them but they will not breed enough birds for a YB team so he must take a round from his racers and, in the case of the hens, birds never raced as old birds are therefore an unknown quantity. They are an unknown quantity as racers and as breeders but then the cocks are also unknown as breeders. This is why accurate and detailed breeding records must be kept; there would then be built up racing and breeding records for the cocks that are going to race in future years and breeding records for the hens that are going to be breeders only. If these breeding records are kept accurately and completely then the situation is transformed. A reminder here, that complete means that all failures, as well as successes, must be recorded. If this is done religiously then the disadvantage can be turned into an advantage and the Widowhood hens become progeny-tested breeders.

'Well Bred'

To return to the pair that we considered mating together, a few pages back, not much will be known about the yearling cock's racing ability, but of breeding potential much more may well be known. The clue to that is its pedigree. A pedigree should always be kept for every bird in the loft, and it should be filled out with care and accuracy. I prefer a pedigree to have on it not only the ring number and sex of the bird, but also some details of its performances as a reminder. In addition full details of every bird's work during the year should be kept in some sort of notebook.

This yearling cock we are considering is well bred: its dam while not being a Federation or Combine winner has flown consistently, and has raced from 500 miles (800 km) on four occasions, being in race time always and being first bird to the loft twice. The parents of the dam include a bird which won the club twice at the 200-mile (320-km) stage and has been flown to the limit of 600 miles (960 km). The other parent has several good performances including first Federation. On the male side of our hen's pedigree, the grandsire was again a bird which has proved itself, having won several good club positions. The other grandparents are similar birds which while not being outstanding have not disgraced themselves. It is not until the great-grandparents are reached that stock pigeons are found which have never been tested on the road although they are obviously able to produce good youngsters.

Stock Birds

To a certain extent every stock pigeon must always be an unknown quantity and even those that have produced good youngsters time and time again may by some quirk produce a bad one. This then, roughly, is the pedigree of our well-bred yearling. It has been given in some detail to show what is meant by being well bred. A bird of this sort, if it can breed successful birds, will make a valuable stock bird. This again shows the value of keeping records. If the youngsters from certain birds or a certain pair are found to be outstanding, that bird or birds should be carefully looked after and not risked in racing. A bird should not be put by unless it has quite definitely proved itself as a breeder. Unproved stock birds are a greater waste in a loft than unproved racers and should not be tolerated.

We have considered two possibilities in mating our birds. The first was mating the best racing cock to the best racing hen. The second possibility was mating a good race-tested old bird with a yearling of good breeding. These choices are possible only if we are firmly set in our mind which are our best birds. The majority of fanciers could not normally pick out one bird as their absolute best, although they might have a favourite. Rather they would say they had half a dozen of which they had high hopes and which would form the basis of their old bird racing team. One successful bird does not make a successful loft; the bird may be a freak, a chance in a million result of a lucky mating. It may be a magnificent bird but it will not found the loft because the same lucky mating is not likely to turn up again more than once in a lifetime and it cannot reproduce its like.

A System Needed

What the fancier needs is some system by which the whole standard of the loft can be raised so that there is not one freak good bird, but a whole loft full of good pigeons which will breed good pigeons. How is this to be done? How is it that when you walk into many good fanciers' lofts you can see row after row of pigeons similar in size, in colouring and quite often in performance? In short, how can a fancier produce a family of pigeons? The answer is by inbreeding. A lot of nonsense has been written and spoken about inbreeding, line-breeding and crossbreeding. Much of what has been written is worthless, not so much because the ideas are bad, but because there has been no agreement at the beginning on what these words mean. To make it clear, here are the definitions which will be used throughout this book. They may not agree with what is written in other books or with what other people use when they are talking. Nevertheless, a moment's comparison of their definitions with these definitions will possibly save much puzzlement and confusion.

Inbreeding and Crossbreeding

Inbreeding is quite a wide term and refers to the mating of any bird with a close relative. Examples are the mating of a bird to its brother or sister, cousin, aunt, uncle, nephew or niece. It extends to the equivalent relationship in pigeons of second cousin and great aunt or great uncle. It can be used for even more distant relationships but this is not normal. Inbreeding includes within its scope line-breeding. Line-breeding is the mating of a pigeon with its father, mother, son, daughter, grandchildren or grandparents; in other words an ancestor or descendant of that particular pigeon.

The other breeding possibility is crossbreeding. This is the introduction of a pigeon in no way related to its mate. The differences can be visualized in the form of a pedigree. An inbred mating is the mating of a bird to any other on that bird's pedigree. It can be any bird above or below or on either side of the bird to be mated. A line-bred mating is the mating of a bird to any bird above or below on the pedigree but not to birds on either side. A crossbred mating is the mating of a bird to another bird that does not appear on that pedigree.

Which to Use?

Having now decided on the meanings of the words we can consider the advantages or disadvantages of these various systems of breeding. Let it be understood at the outset that no one system is better than another. Some are better in certain respects; others give more satisfactory results in other directions. Our aim is a family of first-class racing pigeons capable of producing first-class progeny. The first thing is to blend the birds into a family, what the scientists call securing increased genetic uniformity. This can only be done by inbreeding and when carried out over a number of years this will produce in most lofts a sufficient degree of uniformity.

In some lofts that have recently been assembled from a hodge-podge of pigeons, there is one bought here, a gift bird there, pigeons from twenty different lofts and with twenty different breedings. These will take more time to sort out and in a case like this it may be more than five years before the results are visible. Fanciers with this miscellaneous collection are most in need of scientific breeding methods. A more rapid method for them would be to test their birds hard on the road and then to line-breed to their best, subsequently inbreeding.

Magnification

Inbreeding itself will not produce racers. The result of inbreeding is to magnify the points of a pigeon. It will bring out the good points of a pigeon and, it should hastily be added, it will bring out the bad points. The good points that we wish to encourage are homing ability, the will to race, stamina and resistance to disease. The points we wish to eliminate from the pigeons are obviously the opposite to these. By continued inbreeding we will, therefore, make the distinction between the good and the bad pigeons even more clear. Even if we do not cull the worst in infancy, they will soon fall by the wayside racing and in this way their influence will be removed from the loft.

Inbreeding would then seem to be the ideal way of producing and perfecting a family of pigeons, but it is not as simple as it appears. If inbreeding is continued indefinitely, what before was a negligible factor will become of considerable importance. Even the very best in the loft will begin to develop faults and to deteriorate. This is a fact which has been recognized by experienced fanciers for years. They have realized that while inbreeding can spell success, inbreeding continued too long can spell disaster. The solution is the introduction of a cross of fresh blood from some outside family of pigeons. If good winning blood is introduced into a declining family of pigeons, the results are quite frequently astonishing. It is for this reason too that crossbreeding has had a great vogue and claimed to be the key so success. It can be; but crossbreeding cannot give lasting success even when one pigeon from an inbred family is crossed with another pigeon from another inbred family.

Crossbreeding, we have seen, will not help the fancier much in the long run unless it is practised with discretion and on an inbred family of pigeons. This fact gives a clue to a complete system of breeding that can give continuing satisfactory results by enabling a fancier to use the best parts of two methods. This system will be considered later in the paragraph on heterosis. Before this can be dealt with, there are several other important aspects that must be understood.

Selecting Birds

The problem of selecting birds to use as a cross is very similar to that of buying birds when starting up a loft. Briefly the thing to look for is that the birds come from a family of known working racers which are likely to have inherited those racing qualities. The birds may be old birds which have distinguished themselves by winning high Federation or Combine honours and they may be purchased, usually at a high price, by auction or privately. Normally it will be found cheaper to purchase a pair of squeakers from a well-known fancier from one of his best pairs. Needless to say, even these will not be given away, but it will still prove the cheapest method.

It is of great importance when buying either old or young birds that the birds bought are from an inbred family. We have explained before that it is possible for the best pigeon in the world to breed nothing but duds. Therefore every pigeon purchased must be one which will breed as near its own type as possible. The simplest method of finding out whether the family is inbred is to visit the loft where the bird has been bred. It should be possible to spot a family resemblance in a majority of the birds. In some lofts you can walk into the loft and all the birds are as alike as peas in a pod. In other lofts there may be two or three types of bird but also a fair number of birds of each type. The loft to avoid is one where every bird is different and there seems to be no predominant colour, pattern or build of pigeons.

Strains

A novice who has been reading *The Racing Pigeon* every week may think I am talking through the top of my hat if I say Belgians do not have strains of pigeons. The advertisements are full of names like Janssen, Delbar, Silvere Toye, Wildemeersch, Verheyes, Van Wanroy, etc. But read more closely. Nearly every one of the fanciers is alive and in many cases still racing. If you read still more carefully you can, sometimes, find the origin of that family of pigeons. Silvere Toye for example, a star from the

1980s, bases his family on Stichelbaut and Vanhee but Norbert Norman is the main basis. Norman's birds come from his father with many introductions including Stichelbaut and Vanhee. Vanhee made their family from a number of fanciers including Stichelbaut. Victor Vansalen in his book *Masters of Breeding and Racing* traces back the origin of these families. Toye, Norman and Vanhee all trace their families back to Cattrysse, Stassart and Jules Janssens, the old fancier and no relation to the popular present-day fanciers Janssen Brothers of Arendonk.

I have chosen this example because all the modern fanciers listed are National champions and the matings are with other National champions. The earlier generations do not get a mention, except perhaps Cattrysse. Stassart and Jules Janssens belong to history, yet if you look at British pedigrees these names still appear. Indeed, if Silvere Toye had been English he would probably be calling his birds Stassarts! That is the British tradition and it is wrong because these pioneer fanciers have been dead seventy or eighty years, which in terms of a pigeon's life has to be the equivalent of over 500 years. When many fanciers talk about their pigeons, they talk about their strain without thinking they can be going back a hundred years.

It must be made quite clear that when we have talked of families of pigeons, we have not meant strains. What is a strain? Some fanciers talk of the Gurnay strain, the Hansenne strain, the Logan strain and even the Osman strain. Some fanciers show with pride a bird that they say is pure Logan. What does it mean? Most of the time it means no more than that the bird resembles certain of Logan's birds. Logan wrote, in a notebook I have got, at length on the origin of the birds in his loft. His book includes details of nearly every bird he had. There was a bird bought here, a bird bought there, two or three birds from one Belgian fancier and a handful from another, and even one bred from a Belgian stray. It was not a family of pigeons, but just a group.

Logan began to breed from these birds and out of this miscellaneous bunch, a family type began to evolve. This type was not the only type of bird in his loft but was the one which caught the public fancy. This is the type to which 'Old 86' belongs, the dark-headed blue chequers. He had before his final sale almost made a family of pigeons. Since that time some fanciers who have a bird that can be shown to have Logans in its pedigree or birds descended from Logans or even looking something like the Logan type state that these birds are the Logan strain. Indeed in some cases it is stated that they are pure Logans.

Pure Nonsense

What a meaningless word 'pure' is. If this were true then every bird on that bird's pedigree would have to be either bred in Logan's loft or all the bird's ancestors must have been bred in Logan's loft. Even if it were possible to produce one bird with a pedigree such as this, it could be any colour or shape since Logan himself had such a mixed bunch of birds. Far better to say a bird is of the Logan type than to call a bird a pure Logan. Even this is not really very satisfactory; neither is it very important. It does not matter whether a bird is pure Logan, a pure this or a pure that; it is no good to any fancier unless it is a pure racer or a pure breeder.

This habit of calling pigeons 'pure' has led some fanciers to go to the other extreme and to proclaim that the basket is their pedigree. In many ways this is good since it means that every bird is tested on the road and that none are kept for the precious blood they are said to contain. This 'blood' is in fact the genetic factors of the original founders of the strain. To race and breed birds with no other guide than that of a basket is not quite enough. Even these birds must be inbred if they are to produce a family of winners.

In the chapter 'Making a Start' I shall go into more detail about what to buy, but I think I have said enough to warn novices not to buy famous names, whether strains or families, whether Belgian or British. My grandfather had a simple formula and I can't better it, 'Buy from a winning loft', and that means winning today, not a hundred years ago, not even ten years ago and not winning for someone else. If you go to a fancier who won last year you can often buy birds at reasonable prices. There are often advertisements in *The Racing Pigeon*, particularly for clearance sales. *Squills Year Book* is another good source for a careful reader.

Finally, there are the charity auctions sales at The British National (Old Comrades) Show. Good pigeons are given by good fanciers to help the charity but they do not always reach the price asked for and a smart novice can pick up a bargain. Auction fever is a disease of most bidders at auctions whether of pigeons or not. Set a limit, say £30, and don't go a penny above it even if it means losing the bird. There will be another chance and the worst mistake a novice can make is to buy expensive pigeons before he or she has the expertise to handle them.

It is unfortunately true that not necessarily the best birds racing are the best breeders and it can happen that a bird will be found that will breed winners time and time again. A bird like that should be treasured and not sent to races, where it could be picked off by the casual shooter or struck down by the playful hawk. It is that bird and that bird only that should be treasured; its progeny should be worked in just the same way as any other birds, for they will not necessarily have the power of breeding winners that their parent had. One bird can be put away as a valuable stock bird but it should only be a bird that has proved its worth as a breeder.

Stock pigeons should never be a major part of the loft, particularly if they are purchased old birds that have to be kept prisoner. Prisoners are a nuisance in any loft. There is always a danger that they will escape and cause trouble and also that lack of exercise and lack of air will lead to bad health and possible impotence. Therefore prisoners should only be tolerated as a necessary evil. In most cases, with time and patience, it is possible to train a bird to return to its new loft.

Breaking to a New Loft

This is a problem that most fanciers have to face at one time or another. It is one that requires patience but which can be solved if the proper methods are adopted. The fancier will be causing himself a lot of unnecessary trouble if he tries to break the bird to a new loft while it is separated. The ideal plan is for a pair of birds to be mated, one of which has always flown out of the new loft.

They should be allowed to settle down and rear one pair of youngsters. This is also a precaution in case the fancier should be unlucky enough to lose the bird. One solution, that seems easier than it is, is to cut the flights. If you refer back to the diagram of the pigeon wing, it will be obvious that the primary flights do not have any blood vessels and therefore if you trim the primaries the bird will be unable to fly until the cut feathers have moulted out, by which time it should be used to its new home. Only cut one wing so that if it does try to fly it will go round in circles. It will surprise you to find how far a bird with both wings cut will be able to fly. I would cut only about 2 in (50 mm) off the wing as that will allow it to fly up to its nestbox and yet prevent it leaving the new loft.

If the bird to be broken is a cock, then just after it begins driving before the second round is laid is the time to make the attempt. The attempt will be much more likely to succeed if the bird has been allowed into a cage of some sort so that it can have a good look round the loft, and is sent out hungry. On the day when the first attempt is made, the hen should be shut in the nestbox. The cock will normally perch on the front of his nestbox after a while. When it is seen to be perching there then the doors of the loft are opened and all the birds are allowed out. It is best that the loft on this particular day should be kept short of food so that they will come in quickly, and with luck the cock that the fancier is trying to break in will come to its new loft.

With a hen it is a little more difficult and even more patience must be used. The best time with hens is normally when they have been sitting a few days, but in this case the cock should be removed from the loft before the attempt is started. Even if these methods are adopted, probably at the first time the attempt may not be successful. The bird must be collected as quickly as possible from its old home and brought back. Preferably they should not be fed at their old home and if the bird persists in returning, then on the second or third attempt the previous owner should drive the bird off and see if it will go back to its new home. My grandfather's suggestion was that the birds that persist in returning, in addition to not being fed, should be doused with a bucket of water. There are few birds which will not make at least one attempt to return to their old home.

The critical moment in breaking pigeons is the first time they are let out of the trap, and unless they are really hungry, and sometimes even when they are, the moment that trap is opened they are off at top speed for their old home. The famous story of this is the pigeon which was sold by an American fancier in Boston to a man in Cuba. When it was let out, the bird promptly made the 1,300-mile (2,090-km) journey to its old home. Indeed even youngsters have been known to cover 400 miles (640 km) on their own to return to their original loft. As it is probable that every bird will make at least one attempt to get back home, yearlings and late breds should be chosen that do not have to be moved too far. It makes collection easier and reduces the hazards of the journey.

A bird broken to a new loft can even be trained up the road and a number of successful birds have been raced to a new loft. Broken pigeons, however, are not always successfully cured of their old habit of homing to another place. One London fancier had a bird which was given to him by a friend of his in the East Midlands. This bird was broken and raced quite successfully, but with one snag. South of the Midlands this bird would home to its London loft, but north of the Midlands it would always return to its original home!

The problem of breaking birds is one which will only occur if adult birds are purchased and not kept as prisoners. There is no absolute necessity why this should be done even if a cross is desired. Opportunities of buying first-class adult birds are rare and it is far simpler to purchase youngsters, bred by reputable fanciers, and I shall have more to say on this later.

The Science of Breeding

So far we have considered the problems of mating and breeding without going into the detail of scientific terms. Unfortunately if the more complicated systems of breeding are to be understood then the fancier will have to master some of the basic principles of genetics. The basic unit is the cell. All parts of a pigeon's body, irrespective of their functions, are composed of cells. These may vary in shape and size, but are alike in having a nucleus and an outer part. Both nucleus and the outer part are made of proto-plasm, the basic substance from which all living matter is made. The nucleus of each cell can be seen through a microscope to con-sist of a lot of tiny threads looking rather like strings of beads. Furthermore, in a normal cell these strings of beads seem most times to be doubled, with each string of beads having a similar one lying alongside it.

Chromosomes and Genes

The name given to the string of beads by the scientists is a chro-mosome and each individual bead on the string is called a gene. These genes control every inherited characteristic of the young pigeon and are handed down in various ways from the parents. Each individual cell of a pigeon will have many thousands of genes and the exact purpose of each is still not known. It is known that sometimes several genes operate together to make some characteristic of the pigeon. About the chromosomes more is known; there are in any particular cell (with the exception noted later) eighty chromosome strings. The eighty chromo-somes in a cell are derived half from the male and half from the female, who contribute forty each. The eighty are arranged in forty pairs with a chromosome from the male lying alongside a chromosome from the female. If you have read my earlier books you will realize that there now appear to be more chromosomes than there were in the past! The reason for the change is that sci-entists have decided on a slightly different definition of chromo-some and part chromosome and I have therefore adopted the latest scientific usage.

Allelomorphs

The chromosome strings carry their genes in the same order every time so that a gene which for example controls a certain colour will always lie alongside another gene controlling the same colour. They are said to have the same locus. The genes may not always be identical, for each gene has an opposite. For example there is a gene which, if it is present in one of the pair of chromosomes, will give the bird red feathering. This red gene will have an exact genetic opposite (called an allelomorph) which will give the exact genetic opposite colour to the young pigeon. The allelomorph of the red colour gene is the gene which gives the blue colour and if this blue gene is present in both chromosomes then the youngster will be blue.

Dominant and Recessive

It does not always happen that both genes are identical; for example one parent may give to the youngster the red colour gene and the other parent may give the blue gene. The resultant youngster will not have a colour that is a mixture of red and blue. The pigeon must be either one thing or the other, it cannot normally be something in the middle. This is the basic principle of Mendel, the founder of the modern study of heredity. Of each pair of dissimilar genes there is one which is the dominant gene. If, therefore, in an odd pair, the red gene lies alongside the blue gene the result will always be red since the red is dominant and the blue recessive. The scientific term recessive is used because it helps to indicate that the gene it refers to does not disappear completely but only recedes into the background.

Sperms and Ova

The sex cells are formed by a special process and when they develop, the eighty or forty pairs of chromosomes of the ordinary body cell are split to form only a single group of forty in the sex cell. In other words only one-half of each pair goes into each sex cell. This process is given the name of meiosis. In the male, of course, male sex cells or sperms are formed by meiosis and in the female, egg cells; both are called gametes. When fertilization takes place a sperm unites with the egg cell (called in the singular ovum and in the plural ova). The united gametes are called the zygote and in this one cell is everything that the young bird will inherit from its parents. As the cells unite the chromosomes take up positions with one chromosome lying alongside a similar chromosome from the other parent so forming the pairs of chains. This zygote cell will split and form other cells but each time these split, the cells they produce will contain the full number of chromosomes. This splitting process (mitosis) will continue and some of the cells will begin to take that special shape which enables them to carry out their special duties.

Homozygous, Heterozygous, Genotype and Phenotype

It is unfortunate that the language of genetics contains so many technical words but they are very helpful once they are understood. The words, homozygous and heterozygous are two of the most useful terms. They both refer to the zygote. The one beginning homo- means that the genes in each chromosome pair are the same and the one beginning hetero- means they are different. In this way, instead of saying that a bird is carrying several latent recessive factors, the bird is called heterozygous. The opposite of this is homozygous in which each gene is similar to the gene in the same position on the adjacent chromosome. The homozygous genes are always the same although they may both be dominant or both recessive. It is this sameness that the name indicates. The importance of this is that a bird homozygous for any factor will always breed true for that factor. If the other parent is not homozygous then variation can occur but this will only come from the heterozygous parent.

The scientists have recognized the difference between what a bird is and what it can breed by giving them special names. The genotype describes the genetic qualities of the bird and the phenotype describes the appearance (or physical qualities) of the bird. If the bird is completely homozygous for all factors, the genotype and phenotype are similar, but if the bird is heterozygous then they are different. Just to complete the picture it should be realized that a bird can be homozygous for some factors and yet be heterozygous for others.

When mitosis takes place a cell splits without reducing the number of chromosomes in the cell, therefore each new cell will have the same genes as the parent whatever the composition of the original cell. No matter how long mitosis goes on the genes will remain the same.

In meiosis the pairs of chromosomes form the gametes. The genes on a pair of chromosomes will not necessarily be identical. It is perhaps simpler to think of mitosis as cell multiplication since a gene can be replaced by its exact opposite, its allelomorph. If the genes on a pair of chromosomes are identical then as has been shown the cell is homozygous for those genes. When a pair of homozygous chromosomes split in meiosis then each gamete will have exactly the same genes. In other words all children will inherit exactly the same qualities from that parent. If it were possible for every gene in all the chromosomes to be homozygous then it would mean that all the many thousands of genes must be identical. It will usually happen that some genes are adjacent to their allelomorphs and the cell will therefore be heterozygous. Heterozygous gametes will therefore not be identical and the children can and probably will inherit different qualities from the same parent.

Complete homozygosity has not been achieved nor is complete heterozygosity possible; thus when these terms are used they are relative each other and not absolute. In every case of heterozygous genes, although there are two different genes, one of these is dominant and will control the appearance of the final product, the phenotype as it is called. In the case of the blue–red colouring gene discussed earlier the dominant is red and if the red gene is present the phenotype appearance will be red. This is true whether only one or both genes of the blue–red genetic pair are red. The phenotype is the same although the genetic construction is different. This genetic construction, the genotype, is the thing that the scientific breeder has to discover and use.

Sex Determination

It is in the first cell that it is determined whether the youngster is male or female. If there are eighty chromosomes to the zygote then the youngster is male; if there are only seventy-nine then the youngster will be a female. These seventy-nine, instead of being arranged in forty pairs as for the male, are arranged in thirty-nine pairs with one unpaired chromosome. This odd chromosome, whether single or double, is the sex chromosome; if single it gives a female; if double, a male. The extra chromosome in the male is now known as the Z chromosome, the female equivalent is the W chromosome. In the older days these were called the X and Y chromosomes respectively, but scientists changed the name because in pigeons the X and Y do not have relatively the same size as they do in other animals; the new lettering was therefore adopted to make this distinction more clear.

Sex Linkage

The missing W chromosome of the female gives rise to important genetic effects. The sex chromosome chain contains, as do all others, many genes and among these the sex gene and others including some of those controlling colour. It has been shown that in the normal cell each chromosome is paired and a recessive gene can be carried hidden by a dominant allelomorph so that it will remain latent until released by a suitable mating. Since the female sex chromosome is single no latent genes are carried and a hen must always be homozygous for sex-linked factors.

Turning to a more practical note, when I ring my youngsters, I put the even-numbered rings on the larger chick as it may turn out to be male and the odd number on the smaller, possibly female, chick. I chose this system because the cock has an even number of chromosomes.

Back Matings

Consider the case of a cock and a hen both heterozygous for red. Since they both carry blue, when gametes are formed, on average each will make one red gamete and one blue. When these gametes meet in intercourse the result will be a pair of reds, a pair of blues and two mixed red–blue gamete pairs. All those containing red will produce red phenotype children but the pair of blue gametes will give blue youngsters. Since blue is recessive and can appear only if both genes are blue, the blue zygote must be homozygous for blue. With red, however, only one zygote is homozygous for red and the other two are carrying blue. Which? To find out, all three red phenotype children must be mated back to their parent of opposite sex. The heterozygous youngsters mated to their heterozygous parent will give the same 3:1 ratio for the hidden colour, but the bird homozygous for red will produce only red progeny half homozygous and half heterozygous for red.

The results can be represented diagrammatically using the symbols 'R' for red and 'r' for its recessive allelomorph, blue.

Parents	Rr		Rr	
Gametes	R	r	R	r
Zygotes	RR	Rr	Rr	rr
Back cross parents	RR		Rr	
Gametes	R	R	R	r
Zygotes	RR	RR	Rr	Rr

These charts are a little misleading unless it is realized that the results show the average of the results of several rounds from the mating and not the colour of the first four offspring.

Increasing Homozygosity

If we mate the offspring from the first back cross back to the homozygous parent then by analysing the results we shall see that there is a continued increase in homozygosity. The parents were both heterozygous; the first generation contained two homozygous birds out of four; this generation back crossed produced eight true-breeding homozygous birds out of sixteen. All the offspring of red phenotypes are mated to the red parents known to be homozygous to red. From the results of this mating twelve homozygous red birds will be produced and only four birds heterozygous for red. This method of mating children to parents is, as discussed earlier, one method of inbreeding. It is thus possible to give as a scientific definition of inbreeding that it is a system for increasing homozygosity in a family. It can be shown by similar methods that crossbreeding results in increasing heterozygosity.

In this example the simple red–blue allelomorph has been used since it is a simple easily recognizable genetic unit. But the red colour can be derived from different genes which give a colour almost indistinguishable from the usual red. This red is recessive and can be recognized since the ground colour is usually more of a chocolate colour than the dominant red and the flights are always dark. The problems of colour breeding are many and varied and undoubtedly full of interest but they are not of prime importance to a racing fancier. However, the question of homozygosity is one that the racing fancier cannot ignore. The old argument of inbreeding vs. crossbreeding can now be examined in a new form – 'Is increased homozygosity desirable?'

To answer this the duties of the genes must be considered. Certain genes, many hundreds in all probability, govern a bird's appearance; as a group they can be called the colour factors. Other genes govern performance and yet others health. The colour factors we can ignore but performance factors and health factors are clearly linked to performance. Inbreeding increases homozygosity not only of the good factors but also of bad factors including the hidden recessive health factors. Health factors are like any others and with inbreeding, recessive health factors will be brought to light. Since birds heterozygous for health factors continue in good health, it seems certain that all bad health factors must be recessive. Inbreeding will make some of the birds homozygous for these bad health factors. These birds will be weaklings or will die as embryos. Inbreeding then will eventually cause an increase in mortality.

Continued inbreeding must therefore be avoided but on the other hand continued crossbreeding will only lead to mongrelization. It is a waste of time to try and breed to any system from mongrels. With scientific breeding the phenotype appearance is not so important as the genotype and the genetic constitution of a bird can be found out only by inbreeding if fresh complications are not to be introduced. Once the genotype has been approximately discovered then it becomes possible to breed to take advantage of the useful performance factors.

Additive Genes

So far only the simple plus or minus genetic action has been considered but when dealing with performance factors the problem involves a different sort of gene action. As well as the plus or minus type of gene there is the type of gene which works in a cumulative fashion. Where in the case of a straightforward dominant or recessive a complete change would be made, in the case of additive genes a little bit extra is added on. An example of this is the gene which controls chequering. For the very reason that only small amounts are added each time it is comparatively difficult to isolate the genes that are additive. Unfortunately most of those factors governing performance are, or so it appears, controlled by this sort of gene. No one has yet been able to decide which genes control the ability of a bird to race well. Obviously those that control its physique and its mental powers are among them but in very few of these characteristics are there any clearcut boundaries. Nevertheless, since we know they exist and since we know something of the way they work, we can try to harness them for our own use.

A Breeding Programme

Before that can be done we must be certain that we are dealing with as many known quantities in the stock as possible. We must be certain that if we breed from a pair of birds they will not vary in any tremendous degree from the parents. In other words the birds must be to a certain extent homozygous: that is, breeding true. It is impossible for practical purposes to breed a bird which is completely homozygous but there must be a certain amount of uniformity about the birds and no mongrels amongst them. For that reason the first step for a fancier with a mixed bunch of birds is inbreeding. For the first three or four years, if the birds of that family have not been inbred together before, there will be a considerable increase in homozygosity and the birds will begin to look like a family. The basis of any breeding programme is a family of pigeons; it is only when this stage has been reached that other systems become worthwhile.

The way to produce a family is by inbreeding but if inbreeding is continued too long then the stock will begin to deteriorate and, as we have seen, it will become necessary to introduce fresh blood so as to check the family's deterioration. Then there comes a difficult problem of what to choose as a cross. If any bird which looks good is used it can introduce into the stock several harmful factors which could ruin the chances of success. It is necessary therefore to be fairly certain that the bird will not carry these harmful genes and that they will not be passed on to the stock. In other words mongrel stock must not be chosen but only a bird from another inbred family.

Choosing a Cross

One of the interesting facts that has been noted by breeders throughout the years has been that certain strains or families of birds make particularly good matings. It is said they 'nick' together or, to use the scientific term for nicking, heterosis occurs. If two families nick together then the results of breeding one with another are exceptional performances. In other words,

2 + 2 are made to equal 5. What a pigeon fancier must try and do therefore is to find a family of birds which will nick with his. If he can do this then he will be able to breed birds of a standard above the average. It must not be forgotten that the results of heterosis are not hereditary and that the exceptional results gained from the crossing of two inbred strains that nick together will not be passed on by the crossbred stock.

Controlled Heterosis

The fancier, then, if he is to gain the advantage of controlled heterosis, must be continually introducing crosses into an inbred family at every generation. Is there a method by which he can do this? He must have not only his original inbred family but also free access to a second inbred family. For this it is best that either he makes an arrangement with another fancier to work together or he keeps the two families in his own loft. Each of the two families is kept as pure as possible. These birds are, of course, raced and not just kept for stock, although they will not normally provide the best performers in the loft. This racing is necessary so that inferior performers can be weeded out and only the best kept.

The best birds of each family are mated together so that the inbred lines are crossed for a limited number of birds in. The best birds in the loft will be produced like this but these birds will not be able to breed birds as good as themselves. Therefore the progeny of these crosses should be expected to be a comparative disappointment. A few of the best can be mated back into the inbred families and absorbed there. Since, however, these crossbreds will be the main racing team, if a large number of youngsters from the crossbred racers are mated back then the two inbred families will gradually become one and a new second family must be set up which nicks with the first family. It can be seen that this involves a breeding programme of considerable complexity where great skill must be used in selecting birds.

The greatest difficulty is to find a family which nicks with the original stock and this can be done only by trial and error. First one pair of birds is tried and, if not successful, all trace of their blood must be got out of the loft. Another pair is tried, and so on. Fortunately a fancier will usually be able to know that so-and-so's birds cross well with his and in that way a search for a good family from which heterosis can proceed will be greatly simplified. There is a useful clue provided in the family or so-called strain. Suppose you have had birds from Arthur Beardsmore. If you are looking for a cross you may decide not to go back to the original Beardsmores but instead to look for a fancier who is racing them successfully. They will be just a little distanced from the originals, especially if they have been blended into an existing loft. This is only an example intended to show the way to look. The same would apply for example if you raced Janssens but because there are so many of them, you might do better by going back to the root stock of Janssens. It is not an easy process but if you get it right it will not only put you on top but keep you there.

Preparing for the Races

The whole object for most people in keeping racing pigeons is to race them, and the whole of the loft organization must be built round this. A system of management or, as the Belgians call it, 'a regime' must be decided upon. Every fancier, of course, will have his own system. On small points there may be considerable differences of opinion but if they are all considered together there are many points of similarity. In this chapter will be discussed the basic points and a few of the many variations possible.

Weaning

At the age of 4 or 5 weeks, the birds will be gradually weaned from their parents. They will in any case be eating solid food and should be able to pick up grain on their own account. These birds will be transferred to the young bird loft or to the young bird section of the loft. If all the birds have been mated up at the same time, then all the youngsters should be somewhere near the same age. This, it will be found, is a help in preventing losses. In the first few days the birds should be handled to feel for food in their crops. The time of weaning is also the opportunity to give the youngsters a thorough dusting of insect powder. At the same time they can be examined for possible defects or infections such as canker.

Hand Feeding

If no food can be felt in their crops then rather than return them to their parents they should be hand fed for a day or two. Hand feeding is simple but tedious. If two work on the job it can be quite quick, although it is not really difficult for one person. First a small pot of peas should be soaked in hot water for several hours. Then get the bird and wrap it round tightly with a clean cloth. This is to prevent it wriggling and perhaps spoiling its feathers. Then sit down and hold the wrapped pigeon between the legs. This will hold him firm while with one hand its beak is opened and with the other peas are popped in. The easiest hold for the beak is to hold the lower beak with the thumb and second finger and push the upper bill with the first finger to open the mouth. This technique is not necessary with younger birds still in the nest. They can be fed by just holding the head without removing them from the nestbowl since they do not struggle.

Young Bird Training

After the youngsters have been on their own in the young bird loft for a week or ten days they should be used to their surroundings and they will have spent a fair amount of time sitting round the windows or the trap having a look round. They will have been fed regularly, a light feed in the morning and a full feed at night, each feed accompanied by the rattling of the corn tin as usual. After they have settled in the loft, their morning feed should be missed on a reasonably fine day and the trap opened. They should not be driven out of the trap and the fancier after opening the doors should leave them alone for an hour or so. The birds will begin to walk out, sit on the outside of the trap, sit about on the roof top, and perhaps even have a fly round. After they have been out for an hour or so they should be fed and, as they will be hungry, they should all come in at the rattling of the tin. If bob wires are used, it is not necessary to make them come through the bob wires on this occasion although afterwards they should do. With luck the fancier will find all his birds have returned.

The first adventure out of the loft should not be delayed too late. If it is left for longer than about a fortnight after weaning the fancier will find that as the birds become older, they are more eager to fly and have the necessary strength to fly short distances; they are too strong on the wing. When he opens his loft for the first time, these older birds will fly out with a joyous flap of the wings and by the time they have begun to tire they will be outside the area they know and will not be able to return home.

This is the traditional wisdom but Bob Hutton, a good flier who lives just round the corner from me, uses almost the opposite methods. Because he works late he has time only for the old birds when he gets back in the evening. He keeps all his young birds in until much nearer YB racing. Sometimes there will be first and second round youngsters in together. As usual they are first let out late in the evening but he finds that his losses are not more but less than average. The important thing is that by then all the birds are strong on the wing so there is no danger of the older ones dragging the younger ones too far away.

Local Knowledge

Whatever the homing instinct is, it is undoubtedly true that the last mile or two home depends on the pigeon's knowledge of the neighbourhood. For that reason the birds should fly well round home before they are basketed. If the birds are allowed on to the roof early enough, they will begin to fly when they feel up to it. As their strength increases so will their knowledge of the neighbourhood and their ability to find their way home from even farther afield.

Flying Off

When the youngsters first take off they will fly quite chaotically about the sky in all directions and dart playfully round and round. After a little while they will begin to batch up more in the style of old birds and they will fly round and round in a bunch. They will usually be eager to be out and flying but they will still spend most of their time near the loft. Eventually one day they will be let out and by the time the fancier has turned round they will be nowhere to be seen. After perhaps an hour or two, when the fancier is beginning to wonder what he will be doing for young birds that year, they will come back out of the blue, circle round two or three times and drop on the loft. Probably after this first fly they will look a bit whacked and possibly one or two of the younger and weaker ones will be left behind and may not return until much later, if at all.

There is nothing to worry about in youngsters flying off or running like this; they will all do it and it is most valuable to them. There is, of course, the danger that if birds a week or two younger are mixed up with older birds, these younger birds may be carried off when the birds fly out like this. To prevent this many fanciers breed their first round of youngsters and then put all the birds on pot eggs so that they do not hatch a second round. When the birds have been running like this for several days, then is the time to think about training.

I like the birds to have as much running as possible but it depends on your personal timetable. Work backwards from the first race and allow yourself 3 weeks for training tosses. If the weather is fine for those 3 weeks then there is plenty of time, but if you get a spell of cold or wet weather you will still have enough. In general the more the birds fly off and go running the better because once you start basket tosses they stop running.

On the training of young birds there is a great difference of opinion but the novice fancier can be assured that very few young birds can be over-trained. Some fanciers have their first toss actually at the loft. The birds are not given their morning feed and are put in a basket. A water trough is put on the outside and they are encouraged to drink from it. If necessary their water fount should be taken out of the loft the previous evening so that they should be thirsty enough to look for water in a strange place. If the birds will drink from the trough one hurdle will have been cleared, for young birds must be taught to drink in the basket before they are sent to the first race. They should also spend at least one night in the basket at home so that they will settle down more readily on the first race. Indeed they should be basketed several times during their training just so that the fancier can be certain that they have learnt to drink while in the basket.

The First Toss

On the first basketing the basket is put just outside the loft. When the birds are liberated, they will fly round, and then go back into the loft probably fairly quickly as they will not have been fed before basketing. When this first basketing has been accomplished they can then be taken away for training. Many fanciers make 2 miles (3 km) their first toss followed by a toss at 5 miles (8 km), then 10 miles, (16 km), 15 miles (24 km) and in steps up to the first race point at about 60 miles (96 km). In some urban areas where 2 miles (3 km) would keep the birds still inside the streets, they may be taken as far as 10 and 15 miles (16 and 24 km) for the first toss, and they seem to come through all right. One year the first toss I gave my birds was at 14 miles (22 km) and while they were out a thunderstorm came up suddenly and the whole lot got caught in the pelting rain, thunder and lightning. Nevertheless I was only two short and one of these was reported later.

What can happen if the birds are not taken far enough away is that when they get out of their baskets they may over-fly their lofts and get into strange country without noticing it. The problem of over-flying can also arise when the birds are carried past their lofts by a strong following wind. Many fanciers for this reason give their birds at least one toss, north, south, east and west from the loft. Still others do not believe in a lot of training and just give their birds one toss, mostly so that they can say that they have been trained. Birds handled in this way can and do win short races but only because they are brought over the distance by the rest of the birds and just happen to trap quickly. It is all a matter for experiment which every fancier should try.

Amount of Training

Again there is a marked difference in the time taken over training by fanciers. Some do not start training until a fortnight before racing and then train intensively until the first race. If they happen to strike a week of bad weather then the birds get that much less training. Others begin their training much earlier and take it more slowly, training perhaps on Tuesdays and Fridays only and avoiding Saturdays because of the danger of clashing with race birds. In this way a few days lost because of the bad weather can be made up easily. Birds trained gradually can be taken right up to the first race point and then from some nearer tossing place given a single-up if it is possible.

Direction of Training

In the old days fanciers would train their birds according to the way the railway went, but nowadays railway training is so expensive, even where trains exist, that almost all of it is done by car. A few still use bike or motorbike and because of the cost of petrol these days there is much more club training by road, using the transporters which are otherwise lying idle mid-week. In general people will train the most convenient way. From my own loft in North London I simply go on the motorway, which takes me slightly east of where I would normally go, but as the motorway saves me perhaps an hour or more I just have to make the best of it. If I had more time, perhaps I would travel more closely on the flying route, but time, of course, is the important factor. By road, single-up tossing is comparatively easy – it just needs even more time.

The fanciers who start training fairly early, train up to the first race point then bring their birds back and give them one or two tosses a week at about 25–30 miles (40–48 km) until the first race. The scientists have discovered what is called a 'zone of dislocation' where the two different types of homing instinct that have been discovered do not quite match up, and it lies between 30 and 50 miles (48 and 80 km). I know of many successful fanciers who train from 30 to 50 miles (48 to 80 km) but it suits me not to take any chances. I normally train regularly up to 30 miles (48 km), or thereabouts, where there is a convenient high point with a water tower where I am a regular visitor. If I go farther than that it is nearly to the 50-mile (80-km) mark, but that is usually only with those birds which have missed the early races. It is unfortunately true that if birds are trained thoroughly it is more expensive than just a casual training since the price of petrol makes all long-distance training expensive. A fancier should not pay too much attention to this when considering what plan he will adopt. As soon as he begins to neglect his birds for the sake of saving money he is on the road down. It is important that birds should not be brought back too far after being trained up to the first race point, for the birds do not home so well from distances under 25 miles (40 km) or so once they have been trained. No one seems to know why, but the fact remains that trained and raced birds are more likely to be lost at 10 miles (16 km) than 100 miles (160 km).

Late-Breds

It can be seen that the training to be given to youngsters needs a lot of consideration before any plan is decided upon; even more careful consideration must be given to the training of late-breds. These youngsters bred in July and August or later will not be old enough to race as young birds even if the Federation organizes come-back races. Late-breds are the result of matings of race-tested birds which have been allowed to rear after the old bird season. As I explained earlier some fanciers do not let their long-distance candidates rear before racing so that one or two rounds of late-breds are all that are reared. Peter Virtue, a winner of the Scottish National from Beauvais, says that half his present OB team are late-breds and they out-perform the early-breds and he is talking about birds that come out of the nest in September and October. It just shows that it can be done. These late-breds can be a valuable part of any loft. They are a fancier's safeguard in the event of a disastrous young bird season and can be the progeny of the fancier's best birds without hindering their racing. They need, however, proper treatment. They have to be built up so that they can withstand the winter just as well as the early-bred youngsters. They have to be fed sufficiently well so that an early winter does not catch them under-developed.

They must as youngsters have a certain amount of basket work and provided September is reasonably fine a number of tosses down the road will give them the right amount of confidence that they need. It is important, however, to realize that as late breds they will not moult all their wing flights. The body moult will take place almost as usual, but some wing flights will probably stay with them until their yearling moult. This, of course, will greatly influence how they can be raced as yearlings. The moult is important to every pigeon but with late-breds it is even more important, particularly as irregularities in moulting are not uncommon. The late-bred may drop two flights in each wing at the same time, leaving a nasty gap. In addition these late-breds as yearlings will be unraced. Even if it has been possible to get them down the road as far as the first race point by September they will have had no racing experience unless there is a local after-season short race. Therefore the training given to them at the beginning of the yearling year should be similar to that given to young birds rather than that given to the old ones. In addition to special treatment in training, late-breds may need special treatment when being paired up and this is considered later.

Old Birds

Old bird training is in many ways quite different from that given to the youngsters. Most of the old birds have been over the course many times as youngsters and do not need to be shown the way. Training for them is more reminding them of the way, teaching them short cuts and getting them into the right physical condition for the races. The business of teaching birds short cuts, or as some fanciers call it, a good line, is one on which there is endless discussion. Two fanciers I know in one club were almost invariably only separated by decimals, both near the top of the list and both in their time had won high Federation and Combine honours. Although their lofts were separated only by a few hundred yards, each had an entirely different idea of the line the birds came on, and while both were racing on the North Road one trained his birds to the north-west and the other to the north-east.

For those who have the opportunity to use different liberation points it is well worth trying to get some idea which way the bird comes. As a rule pigeons will go round a hill rather than over it. They will not usually make directly across water, if by following the coastline it is helping them in the general direction of home. In the same way when the wind is easterly they will be blown across their course and usually come in from the west. With a west wind they will come in from the east. It follows then that given a known wind direction pigeons will tend to follow a certain course, but if the wind is in the other half of the compass then they may choose an entirely different route. Using this knowledge a fancier can decide on a number of points which will give his birds experience on both these approaches to the home loft, and by this he will give them the greatest chance of breaking free from the mob and making their own way back to the loft unhampered by the others.

Condition

It must be emphasized that the principal job of old bird training is to get them into condition for the racing season. The long winter months and the long hours on the perch will make the pigeons fat and flabby. Their muscles must be tightened up, their breathing improved, just as a human athlete must go into training before a race. The training that the old birds have to be given will vary, particularly according to the amount of flying they do around home. If the old birds fly well and vigorously round their home loft for half an hour twice a day, training down the road need not be too extensive or too expensive. On the other hand if the pigeons spend about 10 minutes in feeble flapping and gliding round the loft, then plenty of road work will be needed to get them fit. Pigeons vary enormously in the amount of work they will take round home and the problems of getting them to exercise round the loft if they are not willing are not easily solved.

Exercise

We have already discussed, in Chapter Three, the question of Natural birds that will not take exercise, but it is a problem that is one of the most difficult to solve. If flagging has to be adopted it might be thought that this would upset the birds if they were very tame and friendly with their owner. This is not so, because pigeons seem to separate entirely what their owner does outside the loft from what he does inside. The disadvantages of having to flag birds are obvious, since not all fanciers will be able to spend half an hour twice a day keeping their birds on the move. What is worse, a fancier may find that the pigeons get cunning and settle on a roof out of reach and possibly out of sight, waiting until it is time to come in.

Another reason why Natural birds do not fly as long as they should is that quite often they are sent out too hungry. They should be sent out of the loft hungry enough to trap in readily when the tin is shaken and yet not so hungry that they are desperate for food. This is a balance that a fancier has to achieve for himself. If birds have been trained always to come in when the tin is shaken, then after a while they will come in automatically even when they are not hungry. A compromise for fanciers whose birds are neither trapping well nor flying as much as they should round the loft is to fly them twice a day. The first exercise period is early in the morning when they have been without food from the previous evening.

This applies only to birds that are to be raced on the Natural system and for them this morning fly is short, perhaps 10 or 15 minutes and is used largely to train the cocks to trap well. These cocks are lightly fed, if possible separately from the hens which, if this first fly is early enough, will still be sitting. In the later part of the morning when the cocks have changed over, then the hens can be let out for a trapping fly. If it is not possible to have this exercise period in the morning because of work, then the hens should be given a light feed and given their exercise in the after-noon or early evening and fed in. When the cocks come off the eggs they can often be given another fly. Birds should not be chased off eggs to take exercise but no harm is done if they leave the nest of their own accord. The Widowhood fanciers do not have the same problem because obviously there are no hens to go out. Many fanciers have also found that once the cocks are turned over to Widowhood they tend to take exercise more eagerly. In the old days when a man was at work the wife was at home and many became experts with the hens. Today many wives are out at work too so that Widowhood racing and train-ing looks more attractive.

Flying the Young and Old Birds Together

If you are racing Naturally then although normally the young-sters and the old birds will be exercised separately, on some occa-sions it is possible to use one group to help the other. When old birds are being prepared for races they do not always seem to be taking enough exercise, or if they do go out for exercise, do not fly with a will. If this happens it is often advisable to turn out some of the youngsters, which by their youthful enthusiasm encourage the older ones to fly harder. Old ones can also be used in the same way to help the youngsters when they are being exer-cised round home between races.

This is one way in which the old ones can help the youngsters but the most help can be given when the young birds are just beginning to fly round home. Usually the youngsters will start off flying in all directions without any idea of bunching together. If a few old birds are turned out with them they form a core round which the youngsters will group themselves and the whole lot will soon fly together as a kit of youngsters. The useful-ness of old birds does not end with this alone. A fancier will find that when his young birds are flying well around home, there is a danger that they will go running off before the younger ones of his team are old enough. Some fanciers do not exercise their youngsters in the morning because of this but if the old birds are let out with the youngsters usually these old birds will prevent youngsters going on the run. It is, of course, inadvisable to overdo this, since youngsters should go on the run before they are basketed. It is possible, but not advisable, for youngsters that have never been running at all to be trained and raced. I have known fanciers who do not like shutting up the hens all season using them to control the exercise of the young birds but this rather depends on the details of the way the Widowhood cocks are prepared.

Feeding and Breeding

The problem of exercise is not the only thing that the fancier has to contend with, since at the same time as Natural birds are being prepared for the early races some will be breeding youngsters. While they are giving solid food to the squeakers the old birds should never be long without food, or both they and the squeakers will suffer. Here again another compromise is necessary in order that the youngsters may get the best possible attention even if the training for quick trapping is omitted for the time being. If the birds are mated up at the right time they can rear one round, the basis of the young bird team before serious old bird training commences. They are not allowed to lay and hatch again until after the early races. In this respect birds will be treated the same whether flown Natural or on Widowhood. It is only after this that the systems separate. If you are looking to the long-distance races flown Naturally, in many cases this separation is desirable since the bird will be at the right position on the nest for its best race and if a suitable plan like this is adopted then some of these difficulties can be overcome.

The problem of getting the Natural-raced bird in its best position on the nest, as it is often called, is one which will be considered in the next chapter. The basic idea is to give the bird the greatest incentive at the most useful time to get home. Needless to say, if the birds do not know the route and if they are not in condition then any amount of preparation in that respect will be wasted.

Switching

In discussing the preliminary training of young birds reference was made to the fact that some fanciers train their youngsters from all points of the compass whereas others train them only along the line of flight. This is more common among fanciers in the south of England, particularly London fanciers who also fly on the South Road. The London South Road organizations actually begin their training in a south-west direction and gradually work round even more westerly until at their farthest race point they are flying only a little south of west. Several years ago some organizations arranged young bird races from Caen in France, lying nearer due south than any of the previous race points or normal training tosses. The races were bad largely because of the weather but it is worth considering whether the almost 90-degree change of direction had any effect. Some organizations like the East of England Championship Club obviously have found it so for all their race points are in a direct line (or nearly so) with the first French race point.

Many fanciers flying both North and South Road have found that they can take birds raced regularly on one route and race them with success from one of the more distant points in the other direction. Some fanciers even believe that this gives added speed. Without exception this switching has been done with old birds. Among the many experiments performed by Dr Matthews at Cambridge University were experiments on young birds which showed that if trained in one direction these youngsters would tend to fly off in that direction even if the home direction was different. This tendency was almost unnoticeable in old birds but quite remarkable in youngsters. The question this raises is whether or not the London birds released at Caen flew off in their accustomed direction towards the north-east instead of going across the Channel homewards. It is a point on which a fancier must satisfy himself before he attempts switching young birds. It would be interesting to find out how successful switched birds were if released singly.

Single-up

In a single-up release, as explained before, a fancier will let go each bird separately and not release the next until the first is out of sight. If he is able to release them from a high point and follow the birds with binoculars, he can be certain that each bird is well clear before the next one goes. In this way he can make sure that every bird of his is capable of doing that particular journey on its own and he will give it valuable training in breaking from the mob on race days. The training programme should whenever possible include at least one single-up toss. The snag is that there will be a 5–10-minute interval between each bird and even with a small loft the fancier may spend 2 or 3 hours letting them go.

It is risky, although sometimes necessary, to give training tosses at the weekend when huge flocks of race birds can easily confuse those being trained but the dangers are far greater in the case of single-up tosses and it is better to do without them rather than risk a single-up toss at the weekend. All training where possible should be done on weekdays but unfortunately fanciers cannot always be choosers.

The times taken by the birds on single-up tosses are not usually very good since there is a tendency to hang around and wait for the next bird and a lack of the competitive urge. For that reason many fanciers advise double-up tossing in which two birds are tossed simultaneously and race each other to the home loft. This is by no means as rigorous a test as single-up but is almost as good training for breaking from the mob. In both single and double-up tossing it is best to have someone responsible at both ends. If the fancier is unable to have a partner at the home end to check the birds in then he would be well advised to send his birds with a clubmate or someone local who organizes group tosses who will carry out his instructions at a reasonable cost. While single-up tossing is by no means essential it is a great help in the training of those birds that will be required to race from the longest distances and will probably have to finish the journey on their own.

Baskets

It is preferable that a fancier should decide what sort of training programme he intends to carry out before he purchases his baskets. In that way he can make sure that he gets exactly the right sort of equipment. Pigeons can be carried in anything; I have seen them taken away from auction sales in carrier bags and even ordinary paper bags but no one could recommend this. For single birds there is nothing to beat a stout cardboard box specially made for the job. It is a mistake to use a box too large, for the pigeon can get thrown about in it whereas if the box is small there is not much chance of injury. Strangely enough events have changed when we look at one-bird cardboard boxes. Years ago I remember standing on one of the old sort to prove to British Rail how strong they were. Since then Amtrak and Interlink have introduced their service for getting strays back to their owners. What a wonderful boon this has been to the sport but aside from that the fact that they give away free boxes means that nearly every fancier has plenty of one-bird boxes. And it has shown that 24 hours in a box does no harm to a pigeon. For any longer time a basket is far more suitable. Although transporter crates are today made of wood or even fibreglass, I still think there is nothing to beat a basket for comfort. I even made a crate to get my birds used to being crated for the transport but I find I use a basket instinctively every time.

Baskets are normally made from buff willow in various shapes and sizes. The larger sizes were the old club panniers holding thirty birds and the smaller are the eight- to ten-bird baskets in which a fancier carries his birds down to the club. Race baskets made of sheet metal have been tried but have never proved successful although they result in a saving of weight. The baskets I buy now hold about ten to twelve birds. I can carry one in each hand when taking them training and that size fits into the back of a car easily. Unless you plan to change your car soon, it is commonsense to measure the car boot and plan your baskets to fit exactly. It will cost no more if you order in plenty of time and makes life a lot simpler. Only too often the standard sizes are an inch too long or a couple of inches too wide for the space available. If asked in their quiet season, most basketmakers will be quite willing to make a basket to your requirements. The ideal places to buy baskets are at the national shows, like the British National (Old Comrades) Show, where the various manufacturers have their goods on display. It is always best to buy the baskets just a little larger than required since overcrowding in a small basket can lead to broken flights or even the injury of a bird. Another type of basket which is very useful, particularly for those Natural racers who want to keep their birds separated in the baskets, is the partition type with a double flap on the top and a partition across the middle so that cocks can be kept on one side and hens on the other.

Widowhood fanciers will often choose something more like the partition show basket where each bird has a separate compartment and the sides are enclosed. They take up more room than ordinary baskets so I prefer the Belgian style with split cane or aluminium sides because they are lighter and take up less room in the car.

All wicker baskets should be cloth-lined to reduce the danger of injuring a bird on a broken twig or of it getting a foot caught. Many fanciers recommend having the tops of training baskets solidly woven over to prevent dirt falling in, whereas others prefer the open type of club basket. These are made with open spaces through which the birds can be seen and fed if necessary. The position of the spaces on the sides of a basket are very important since water troughs have to be fitted on to these. If the spaces are fixed too high or too low then the birds will not be able to drink comfortably and if the wicker work is not well finished off it is difficult to fit the troughs on easily. If these side spaces are too low it is also possible for the water to be fouled from the inside. The disadvantage of the Widowhood style of basket is obviously that birds cannot be fed or watered in them.

Birds should never be put in a basket until the bottom has been covered with wood chips. Sawdust should not be used as it can get into the bird's eyes and mouth and also because it often contains wood splinters. Hay should never be used because it has been shown that hay and straw can cause various diseases to spread and its use is illegal. For a while it was quite difficult to buy wood chips but recently they have become available in bales. Each bale is hard packed and will last more than one season. I suspect these chips are not industrial waste but specially made for the job. In our Federation we used peat on corrugated board to line the crates. I tried peat in my baskets but found it became too dirty and in any case we are now urged not to use peat in our gardens for ecological reasons so I will happily stick with wood chips.

Training Weather

While training tosses are being made the fancier should be most careful about the weather, even on a short toss. A little bad weather can ruin the fancier's plans for the racing season if he is not careful. The three most important things that affect a pigeon when it is flying are the wind, the rain and fog. Of these the last is by far the greatest enemy. Rain showers will not normally affect the birds but in heavy rain they will come down to shelter and start again when it has passed. If they have been racing hard it is quite possible for them to get severely chilled in a storm and sickness can set in later.

Wind

The wind is very rarely too strong for pigeons to fly but it makes a considerable difference in the amount of wear and tear on the pigeon's frame during the race. With a wind behind it a pigeon will come home looking as though it had not been a yard but with a stiff head wind blowing the pigeon will be worn out and can indeed on some occasions become unsuitable for further racing that season. Depending on how a bird returns from a race, so must a fancier's treatment vary during the next week. A pigeon that has had a hard race needs rest and good food to replace lost energy stores and to renew worn tissue. Quite often after a hard race the fancier will rest his candidate for one or even two weeks before racing him again. This does not mean that the bird must not leave the loft box for it should certainly receive its proper amount of exercise round the loft or in training tosses.

Racing Weather

The study of the weather is particularly important for those who share the responsibility with the convoyer for giving the order to liberate. They should have more than a slight knowledge of weather forecasting and of weather forecasts issued from Meteorological Offices in this country. The convoyer should be the key person for collecting this information. He wants to know not what the weather is at various places on the route but what the weather will be at those places when the birds reach them. For this reason line of flight telephone messages from fanciers lying on the route over which the birds are going to fly have very little value unless the fancier the convoyer contacts has some knowledge of weather forecasting. Specialist forecasts are available for convoyers but for general use I think the radio or television forecasts are very widely used. The early evening TV forecast is familiar to most fanciers and is usually enough for them to decide whether to train the next day, and many use them on marking night to find out when the birds will go up on race days. They are very useful but do not replace specially prepared ones.

These detailed weather forecasts concerning the next twelve hours are rarely seriously wrong. Occasionally unexpected bad weather may develop even in this short time but this is unlikely. It is essential that the forecasts should be completely understood. On one occasion when several organizations were racing from France, the forecast indicated north-east wind and fog patches in the Channel. The combination of the two proved overwhelming and the races were disastrous. One of the most difficult things to forecast is a shower because showers are always local and while there may be showers of rain in one place, a short distance away the sun will be shining. Nevertheless, providing the forecast is properly used it can help the fancier greatly. In the same way, when a fancier is preparing his birds by training, he should never send them away without first getting a forecast, whether it is

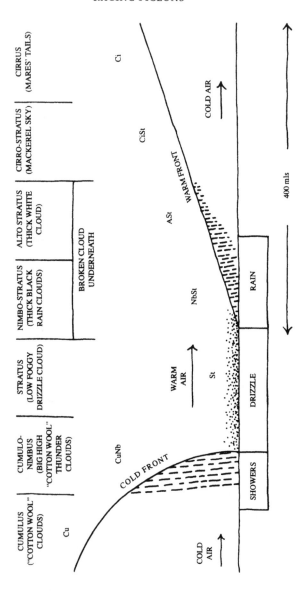

19 WEATHER ASSOCIATED WITH PASSING FRONTS.

168

from radio, television weather map, the newspapers or by telephone. Remember always to get the latest forecast as forecasts can change from hour to hour.

Summer Storms

It is worth noting that with the fine summer weather that accompanies anticyclones there can be thunderstorms. These thunderstorms build up during the day, so in the early morning there is very little danger from them. For this reason birds should be given their training tosses as early in the morning as possible in this sort of weather. For the fancier who has to go out to work or to business during the day this means he must get up that much earlier every day. If he does this, his birds should be home before the storms have built up. He must before he leaves make arrangements for them to be able to get into the loft, or arrange for someone to let them in. Better still, he must be up early enough to see them arrive himself.

Frontal Weather

The thunderstorms that accompany the long periods of fine weather are not the only type of bad weather that birds have to encounter; worse than these are the cold fronts centred on a depression. A depression is not necessarily a bringer of bad weather, although a deep depression usually gives very strong winds up to gale force. Depressions are usually formed from and accompanied by masses of cold and warm air. Each of these huge air masses circles around the depression and the cold air mass slowly catches up with the warmer air. The narrow area where the two meet is known as the cold front and it is here that the bad weather occurs. A diagrammatic section through the cold front shows the type of weather and the order in which it may be expected. The cold front is frequently preceded by a warm front and as the cold front comes up from behind and the depression begins to fill in, the two fronts occlude and by mixing together lose their strength.

Winning Systems

Any of the readers of this book who turn first to this chapter hoping to find in it the secrets of success will be disappointed. No amount of writing by me, and no amount of reading by them, can make a successful pigeon fancier. What a book like this can hope to do is to put a reader on the right lines so that he can decide for himself what will bring success.

It would in any case be impossible to give each reader detailed advice without knowing the pigeons which he intends to race since, as has been mentioned so often, pigeons must be treated as individual birds, and in many respects given individual management. Certain treatment and certain handling will make the best of one bird but it must never be forgotten that that same treatment can ruin another. All this chapter can do is to give the reader some idea of the variety and limits of winning methods. I have been fortunate enough in the last few years to help in the preparation of *Squills Year Book* and until recently the *Racing Pigeon Pictorial* in which there are articles written by, or about, the most successful fanciers of each year. In these articles they give information on how they prepare their birds for the races in which they are successful.

In recent years there has been a rise in sprint races and a number of fanciers specializing in them. I don't want to knock anyone's pleasure but I remain an out and out distance racer in my heart. Widowhood will win sprint races and sometimes in skilled hands the longer ones but the majority still stick to the Natural system for the races over 450 miles (725 km). Indeed, as I shall explain later, many of the Belgian Widowhood cracks fly all the shorter races Widowhood but then allow the birds to go to nest for the final races, flying the hens and often rearing a round of late-breds into the bargain. Natural racing is far from dead but in any case to fly Widowhood well you have to understand the Natural system. So even if you never intend to race on the Natural system you should read the next few pages carefully.

Natural Incentives

In the Natural system the birds are kept paired, sitting, or rearing youngsters, and according to the position on the nest that suits them best so they show the greatest desire and ability to return home. It is usually assumed that this is a question of incentives and that a bird which is, for example, sitting 10 days if that is its best position, is more eager to return home than at any other time. Indeed it is not unknown for birds to be given simulated incentives to assist them. A bird not quite due to hatch may have a youngster slipped under her just before basketing, or in olden times an egg with a moving worm inside was given so that the movement of a live embryo was imitated. Recently an electronic egg has been invented; battery operated, it stimulates the movement inside the egg very convincingly. Unfortunately such eggs are quite expensive but I have no doubt that in time the price will come down and they will come into general use with Natural fanciers. Fanciers who do this assume that it is purely a question of the incentive and that the bird races best because of its anxiety to get home. This is not quite the whole truth and with our increasing knowledge of the functions of the endocrine glands, we may realize something more of the reasons for the differences of performance at different positions of the nest.

Glands and Form

The endocrine glands react to many different stimuli. The pigeon, for example, feels the movement in the egg of the youngster hatching; this movement is transmitted to the bird's brain and from there to the pituitary gland. The pituitary gland in turn stimulates another gland which produces the pigeon's milk. In this way, when the bird hatches out it has its necessary feed ready. At the same time as the pituitary gland is stimulating the milk gland there is also a secondary action in which other glands are stimulated, including the adrenal gland and the thyroid. It is quite possible that it is only when these glands are functioning at their peak that the pigeon is really on form.

This indeed is a possible explanation of why some pigeons do so well feeding a small youngster, a time when it would be thought that the greatest demands on their energy are being made. In the same way a different pigeon may produce the hormones necessary for the best performances from different stimuli and although we can only guess at the hormones involved, we can at least discover these stimuli. The Natural system probably depends not only on the simple incentive of anxiety but also on the more complicated hormone reactions dependent on the normal breeding cycle.

Position on the Nest

A study of a few years' *Squills Year Book* winners shows that a bird will win in almost any position, except hens when they are laying. The following table gives some idea of the Natural racing positions first-class fanciers regard as the most suitable:

Driving cocks	2 recommendations
Sitting 4 days	2 recommendations
Sitting 6 days	1 recommendation
Sitting 8 days	1 recommendation
Sitting 10 days	7 recommendations
Sitting 12 days	4 recommendations

Sitting 14 days	3 recommendations
Hatching	2 recommendations
Feeding 1–5-day-old youngster	3 recommendations
Feeding 5–10-day-old youngster	3 recommendations
Feeding big youngsters	1 recommendation

The last position of the nest, of course, corresponds to the beginning of the sitting cycle of the next round. The table does not include those fanciers who state that some birds win at any position of the nest (except laying hens), of which there are a considerable number.

It should be noticed that by far the most popular condition is sitting 10–14 days, a condition in which half the winners of our big national races were sent to races. What is interesting, in view of the comments on the endocrine glands, is the number of those feeding youngsters at the time of basketing.

These figures are by no means in agreement with those of other people who have studied the subject. The late Major Andrew Neilson-Hutton, the well-known author of pigeon books, studied this question, approaching it from a different angle and working on the records of J. T. Clark, the successful Windermere fancier, and he came to very different conclusions. On his analysis, 1 in 7 of the birds sent at hatching time were successful, whereas at the other end of the scale 1 in 25 sent sitting 7–12 days were successful.

The differences between the two sets of figures must be considered since they are important, but fortunately they are not necessarily contradictory, although the period of sitting 10–12 days is given first place in my table, and last in his. The differences in some measure can be accounted for by the general acceptance of the belief that 10 days sitting is the best time to send a bird. This means in many cases that the bird perhaps not quite at its peak may be sent simply because it is sitting 10 days. However, the average fancier looking at a bird due to hatch would think it unwise to send the bird unless something in the bird's appearance and behaviour in the loft convinces him that this bird must be sent because of its fine condition. Putting it even more briefly, only the best are sent when hatching and sometimes junk is sent sitting 10 days. Another difference which may account for a disparity in the results is that his table was drawn up from the records of one fancier, whereas the table reproduced above is derived from many fanciers' notes.

The Best Times

How can a fancier decide which positions are best for his birds? This is a major problem for the novice and when he has solved the problem he will be a novice no longer. The answer, to repeat the thought so often expressed before, is by observation of the individual. Training tosses help, because as the birds come in from the longer training tosses, those that shine can be picked out. The hen that comes in ahead of the mob and goes straight to her eggs without feeding is obviously just right, although whether she will be on the following race day is a different matter. The cock that resists any attempt to move him and look at the eggs when he is sitting on them is obviously another likely candidate. Sometimes a driving cock and sometimes a hen sitting particularly tight will show themselves to be potential winners. The problem can be answered only by observation of the individual, and needless to say the observation must be recorded. The order of birds arriving from a training toss should be recorded whenever possible, and by relating their performance during training and racing, some idea of the rise and fall of a pigeon according to its form can be gained.

The Worst Times

The novice will need to exercise a little care in deciding in what position to send his birds, since it is possible to handicap a bird by sending it in the wrong position. A hen should never be basketed when she is about to lay. This is easy enough to tell if it is a second egg which is due to be laid, but a novice may have some difficulty over the first egg. It will often be laid five days after the cock begins to drive hard, but this again is a matter of observation. In the short races it is possible to feel the vent bones before basketing. Shortly before the egg is laid these will begin to move apart to allow the egg to pass between them, and if a fancier is used to the feel of his birds then he will be able to detect this movement and sometimes even what appears to be an egg in formation. The hen that is 'eggy' like this obviously must not be sent, since she will rest somewhere to lay the egg and even after laying will not be in the best condition to continue racing.

Driving is a position which can prove either highly successful or fatal. Some driving cocks are very successful, particularly on the shorter races, and when the hens are due to lay, then the cocks will quite often be in their best position. However, some cocks, particularly yearlings, drive so furiously that they do on occasion lose themselves by their over-anxiety. One of the dangers is that a driving cock may arrive at the loft at the same time as a hen, and particularly if the hen is 'rank' and eager to be courted, they will spend hours on the top of the loft amusing themselves, but not amusing the anxious fancier waiting to time in a winner.

As has been mentioned earlier, most fanciers reckon that it is undesirable to send a bird with pigeon's milk in the crop, since while it is away this cannot be given to the youngsters and may go sour, discomforting the bird. This, as has been indicated before, is an arguable point. Many successful fanciers find that this is their most successful racing position. It is one, however, that I think the novice would do well to avoid.

Widowhood

In Belgium 90 per cent of the fanciers win on Widowhood or, as it is called there, *veuvage*. In English the system is, strictly speaking, incorrectly named and it should really be called 'Widowerhood', since it is the cocks that play a principal part in this system. There has been a boom in Widowhood in Britain recently but curiously enough in Belgium more and more fanciers concentrating on the Nationals and Internationals are flying the Natural system. Philosophers in this sport may wonder if the fall in numbers of

Belgian fanciers (over 50 per cent) has not been a result of the concentration on Widowhood sprint racing with its greater emphasis on money rather than sport. This has not happened in Britain yet but some see ominous signs in the fact that although the numbers of GB rings sold is going up there is a slow decline in the number of fanciers. The Roundabout system is a form of Widowhood that allows both hens and cocks to be raced and looks to become an important influence in the future. For the present it is time to concentrate on the racing of cocks on Widowhood.

Again, in this system, there is the question of incentives and in Widowhood the incentive which is exploited is the desire of the cock to rejoin his hen at the earliest opportunity. As in the Natural system, this incentive is by no means all of the system, for as the sexes are kept separate nearly all the week the energy of the cocks is conserved and not dissipated on continual courtship. It is my own belief that the almost monastic seclusion in which the cocks are kept contributes as much to the success of this system as the sex incentive.

Glands Again

It is worth noting also that one fairly common form of Widowhood includes showing the cock his hen just before basketing. Although primarily intended to increase the incentive of the racing cock, this stimulation working through the endocrine glands also stimulates racing ability.

Just as there is no single system which can be called the Natural system, so Widowhood can be practised in many varied ways. A few years ago, I studied carefully the best Continental methods of Widowhood, incorporating them in a book, and it was this that made me realize the wide diversity of methods possible.

Separation

The basis of the Widowhood system is that early in training the cock is separated from the hen and only allowed to be with her for a short length of time after trapping. Some Belgians favour the removal of the cock almost immediately after his arrival. Others consider this too frustrating to the cock and do not separate the pair again until perhaps the next day. Some, as has been said before, consider it an advantage to show the hen to the cock on basketing to remind him of her existence. Others consider that this will over-excite the cock, particularly if the system is practised on yearlings. It is on these points that each fancier must make up his own mind, having regard to his own birds and the other parts of his system of management. The popular press (and some fanciers) see the incentive entirely as the male sexual drive which I think is wrong. For example, some Widowhood fanciers do not show the hen to the cock before the race but only turn the nestbowl the right way up in the cock's nestbox. I think in scientific terms this would be regarded as a Pavlovian signal but whatever it is, it is enough to set the cock up for racing.

Loft Alterations

To be practised properly Widowhood needs some alterations in a loft which has been used for the Natural system. These can be extremely complicated or extremely simple. The Belgians, who work this system in a big way, quite often have a completely separate loft built some distance from their main racing loft, in which the hens are kept during the Widowhood period. This same loft may also be used to house stock birds. While this is the most certain method of securing the complete separation of the sexes, it is expensive and increases the work of management.

I have seen in this country small lofts built at the side of the main lofts, and even nestboxes fixed up on the outside rear wall of the main loft for keeping the hens, and it is quite possible just to partition off a section in the loft. This partition must be solid so that the birds cannot see through to the other side, and if possible, it should be sound-absorbent, so that the cooing cannot be heard on the other side. This can be done most easily by making a double partition with an extension beyond the windows at the front. It is, however, remarkable that many successful Widowhood men say that sound is unimportant compared with sight. As I explained earlier in the chapter on building a loft, the sections into which a loft is divided may have to be adapted to accommodate the larger Widowhood nestboxes.

Nestboxes

These are some of the modifications which a fancier must make; the others are to his nestboxes. H. S. Pearson, the famous Yorkshire pioneer in the 1920s, used ordinary nestboxes with a pair of bob wires fitted on the front, rather like the trap nest used

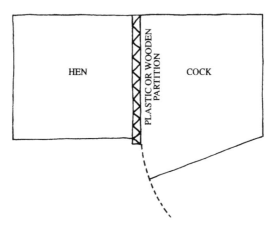

20 WIDOWHOOD NESTBOXES.

for chickens. A more usual type of nestbox is the Belgian type where there is a central partition which can be moved, allowing the cock to be separated from the hen in the nestbox. A typical design is that shown, although variations of this pattern are quite common. In the inner section the hen is kept awaiting the cock's return. When the cock arrives home from training tosses or races, he lands in the outer section where he can be caught, and later, after the rubber ring has been removed, the door is opened allowing the cock to enter the hen's part of the nestbox. By the use of a sliding partition it is not difficult for ordinary nestboxes to be converted to Widowhood.

Reasons of cost have virtually eliminated most fronts other than the plastic ones described earlier and it is for that reason that I suggested making the nestbox size fit the fronts rather than the other way about. Unless you are a handyman with time on your hands, you may even want to go one stage further and buy sets of nestboxes and nest fronts made as a unit. They are made in standard sizes, so again you may have to adapt your loft to fit the nestboxes. Because they are made with chipboard, the prices are quite reasonable. Obviously chipboard is not the same as wood but it does the job reasonably well.

Difficulties, Real

The adaptation of the loft and its fittings is not the only difficulty facing the fancier trying Widowhood. It is obvious from the outline of the system that once the birds are separated and training is started, then breeding is impossible. One solution is to breed very early in January and then separate the birds again to hold back the moult of the OBs. These early-bred youngsters are at an advantage for YB races as they will be through the juvenile moult and will be ready for pairing up. Widowhood fliers are also usually keen YB fliers for the obvious reason that this is the only time the hens are tested. It is also true that a Widowhood sprint racer is going to enjoy YB sprint races.

Next year's youngsters must therefore be bred early before the Widowhood training commences. In other cases, the early training tosses of the year are given while youngsters are still being fed. More important than the prevention of breeding is the fact that the hens are not racing. This means that it is impossible to select the best racing hens for mating, since in many cases their ability on the road is not known. While their breeding ability can

179

be gauged, as it should be, from the progeny, it means that all the hens are in fact stock birds and that for every one raced two must be kept and fed. There is one solution that has found greater favour in Britain than in Belgium and that is the Roundabout system. Racing hens on Widowhood is not new. Dr Tresidder wrote about using it in the 1930s calling it the Celibacy system, but in recent times the name of Geoff Kirkland has been particularly associated with the Roundabout system, which includes the racing of hens widowed. As a system the Roundabout is not yet widely practised but the Widowhood of cocks seems to be increasing all the time.

It is not without its problems. The Belgians consider also that yearlings need very careful handling on this system, otherwise, like driving cocks, their over-enthusiasm can bring disaster. For this reason some Belgians race their yearlings, both cocks and hens, on the Natural system. In this way they can decide which hens they wish to keep. It is also true that the enforced inactivity of the hens can impair their health unless care is taken in their feeding and exercise to counteract this. An aviary for the hens is a familiar feature of Widowhood establishments.

Difficulties, Fancied

The real difficulties of Widowhood are bad enough, but in addition some fanciers invent a few more. It is only too often said that Widowhood requires more time and work than the Natural system. This is not in point of fact true. The initial alterations of the loft require time but after that there is very little extra. The way it is practised by most fanciers, some extra hours do have to be put in but less time is required during the week and more at the weekend compared with the Natural system. The argument, of course, is misleading because the fancier who begins counting the hours he spends in the loft is counting himself out of competitive racing. As the wives of many successful fanciers know, they cannot spend too long in the loft!

Other criticisms which are levelled against the system are that the cocks go off their feed, that they are unwilling to fly and that

Widowed cocks cannot successfully compete over 350 miles (560 km). These are some of the dangerous half-truths which some fanciers especially Natural racers want to believe and therefore do believe, but there have been many Widowhood cocks winning at 500 miles (800 km), there have been many that exercise readily and there have been many that eat quite happily. The poor feeders and the poor exercisers would probably be just as bad on the Natural system.

500-Milers

The question of Widowhood cocks at 500 miles (800 km) is a more difficult matter. Many fanciers flying the Natural system prefer hens for the longer races because of their greater stamina and staying power. Whether this is really justified is another matter but so strong is the belief that many fanciers who practise Widowhood, Belgians amongst them, have a few hens which they keep entirely for the longest races. For the small fancier this means even more complications to his system of management. Widowhood cocks can win at 500 miles (800 km) as O. I. Wood showed. This was in the 1930s when Widowhood was in its infancy. Obviously he did not know that his successes were impossible. Sixty years later the Pau National was won at 560 miles (nearly 900 km) by Harding Bros of Bath with a Widowhood cock. The same year, 1994, the Pau race of the London and South East Classic Club was won on the day at 546 miles (879 km) by Doug Gatland. I have already mentioned the Belgian Barcelona winner who must fly over 700 miles (1120 km) and is therefore like NFC Pau a two-day race but note that none of these birds flew the whole programme before the long race.

The Advantages

The novice considering the problems may well now be wondering why anybody ever bothers to fly the Widowhood method. The number one answer is that in Belgium where the Widowhood system is practised almost exclusively, the

Widowhood fanciers wipe the floor with the Natural fanciers time and time again. Another important feature in its favour is in the question of condition. Widowhood, even though half the pigeons are 'passengers', allows the fancier to enter a smaller team with a greater hope of success. On the Natural system the pigeon will be brought into form for a week or two and then it will have passed its peak and begun to fall off in performance until it can be brought back again to a fresh peak. A Widowhood pigeon is brought into form for the early races and it will stay in that form for perhaps six or seven weeks without any falling off in between. That pigeon can be sent again and again to the races, when Natural pigeons should by rights have been rested. Those fanciers with a small team can by this system compete in the averages against much larger Natural teams. A third advantage that must never be overlooked, except by those who are able to have someone to clean out for them, is that birds kept on the Widowhood system are much cleaner in the loft and their droppings are much easier to clear up.

The Roundabout System

The Roundabout means that the birds go out of one door and come in by another. The birds are separated so that when the cocks are let out for exercise the doors are closed and the hens go into the section where the cocks were. At the end of the exercise the cocks trap into the now empty section previously occupied by the hens. The next time the hens go out and the cocks are moved across. So the Roundabout goes on all the time winter and summer. In my own loft in winter I hang a set of box perches in front of the nestboxes so that it does not matter where the birds settle. They will be given their boxes when they are paired up.

The time you pair up will depend on how seriously you take young bird racing. This is just the same however you fly. With Widowhood it is common to pair up late December for eggs early January. Some Natural fliers do this as well but not many. The idea of the early-bred youngsters is that they will be through the juvenile moult before they go to the races and thus have a better wing than their later-bred competitors. They are also more likely to pair up and so can be raced on eggs or even youngsters. The YB race specialists insist that it does not harm youngsters to be paired as the birds know when they are ready to pair and this includes laying by the young hens.

Many Natural fliers who are used to playing the waiting game prefer not to pair up until February so that the YBs have a more natural moult as they see it. As I said earlier it depends on how seriously you take YB racing. The pressure is on the Widowhood flier and to a smaller extent on the Roundabout flier because they need to know how well the hens will perform. With Widowhood the hens will be raced only as YBs, so racing them paired makes sense because it gives a better idea of what they might have done if they have been raced as OBs.

The beauty of the Roundabout system is that it has most of the advantages of Widowhood and most of the advantages of the Natural system. It is not the miracle system unfortunately because there are some disadvantages. If you pair up early then as with Widowhood you separate as soon as the early round is reared. This is around mid-January and a second round of eggs may have been laid, but it seems that because of the earliness in the year and because the eggs are not sat for more than a few days the OB moult does not start as it would later in the season. Some fanciers only let the cock feed the youngster after ten days and take away the hen. This obviously works best if the early round is single-reared and I know many fanciers are almost superstitious about single rearing, fearing that the egg they replace with a pot egg may be the future champion that never will be. There is no way out of this dilemma. Unless you have feeders in a stock loft you have to take a chance.

When the Roundabout birds are re-paired in late February or early March it is much easier, the cocks and hens go in the nest-box as usual either side of the central position. It will be only hours, possibly minutes before the pair are ready for each other. Some fanciers just let them run together but I prefer the safer way because it also reminds them of the central partition that is going to be such a part of their life for the rest of the season. So far the

Roundabout has been the same as Widowhood. Now start the differences. Under conventional Widowhood the birds pair up, the hen lays and after about 10 days she is removed and the cock sits on his own. He will usually give up after 4–5 days but if he doesn't the eggs are removed.

On the Roundabout I don't think it is necessary to let the hen lay again. They need spend only one day together before training starts for both cock and hen. While they were separated they have been flying in and out of the loft on the winter Roundabout or at least they were on the fine days. When rearing the first round they were flying in and out as if they were Natural birds and the hen section was empty. When they were separated after the first round they went back to the Roundabout alternate flights. In my case in the winter they were flying to the temporary box perches but after the first round they were flying to the nestboxes. I leave only half the box open. The closed half is where the nestbowl usually is but I remove the nestbowl after the first round.

By now it should be obvious that since the hens do not have to lay again they can be re-paired a week later than Widowers and, depending on the first race, this can be in March or even early April because if the birds were raced the previous year and have been exercising well they only need a few refresher tosses. There is an alternative method in which the Roundabout hens lay again exactly like Widowhood hens, but I prefer not to let them lay. Now they are re-paired they continue on the Roundabout. They are together only for minutes or perhaps an hour or so before the hen is removed to the perch section, and the cocks have the nestbox, or more exactly the open half, to themselves. They keep this until the normal exercise period and then they are moved over to the perch section after the hens are let out. If the cocks go out morning and evening and the hens at midday then the cock will have the nestbox each night.

At the early stage I do not think it is critical which stays overnight in the half nestbox. The hens are unlikely to lay and, provided there is only one box for each pair, two hens are unlikely to pair up. With this possibility in mind during the week or two of training I think it is a good idea to use the race procedure at least once a week to keep up the interest of the pairs in each other. If it can be planned in advance to match the feeding it is an advantage but during training it is enough to feed a strong mixture before the toss and depurative after.

By the time of the first race the pair should have learned the rules of the game. On the day of basketing the hens should be in the nestbox section. I don't think it is crucial if, because of the weather, they are not there and the cocks are still in possession. This is one of the beauties of the Roundabout system, it is actually more flexible than conventional Widowhood. If the hens are in the nestbox section then in the afternoon the centre partition is removed and the nestbowl is put in. I have not tried it but if the fancier is at work all day the nestbowl could be put there in the morning but I feel the afternoon would be better certainly with the older hens. Before basketing the cocks are let out to fly and when they trap they find the hens waiting for them. If the weather is bad at this time then the partition door is opened to let the cocks in. They spend 15–20 minutes together in the whole nestbox and with the nestbowls the right way up and then they are basketed.

Having talked with fanciers who are using the Roundabout system in one way or another, it is here that I find most disagreement. Some basket both cock and hen which to me is not true Roundabout but nearer Double Widowhood but it seems to work for them. On true Roundabout only one sex is basketed and that is the one that occupied the nestboxes overnight because as so many Widowhood fanciers have discovered the attraction to the racer is the nestbox and the nestbowl as much as the opposite sex. In the example listed above it means the hens are basketed and of course all the hens should be included. If any look doubtful and should not be sent they must be removed with the race birds and left in the perch section. The cocks are left in possession of the nestboxes until the hens return the next day. If you have to leave any cocks at home then they must be shut in their own boxes with pots of food and water, not allowed to roam the loft, otherwise they may decide they like another cock's nestbox more than their own.

The maestro of long-distance Widowhood racing is Geoff Kirkland and he makes the distinction with sprint Widowhood in that sprinters must always be fed in their nestboxes, whereas distance birds can feed communally on the floor. This is OK but does not apply when birds are away at a race. On their return the usual Widowhood rules apply – leave the pair together for a short time and then the hens are returned to the perch section. It must be the hens because otherwise the nestbowls even without the cocks may encourage them to lay. The feeding for Roundabout birds is the same as for Widowhood: break down after the race; build up before the race.

There is one point here where I differ from many conventional Widowhood fliers. I like to put welcome on the mat when the birds return. This means not only having their mates ready and willing when they return but also giving them their favourite dinner. Some fanciers, and I cannot deny they are successful, feed a depurative mixture or even straight barley to the bird when it returns so that from the moment it traps it is on a break-down diet. I don't think this is much of a welcome. I have yet to find a pigeon that prefers barley as a diet so I would welcome it with its favourite food be it maize, small seeds or peanuts. It does not have to be a lot and then I would switch to depurative. I just do not believe that it would make that much difference to the system and I think the bird deserves a welcome morsel.

How long can a fancier race on Roundabout? This is a problem and I go back to Geoff Kirkland because not only has he more experience with Roundabout than anyone but because he is a keen and original thinker, willing to experiment. For those who don't know, he has also won six or seven Nationals over 500 miles (800 km) if you include the Midlands National and the British Barcelona Club. His view is that Roundabout will work for only six or seven weeks before the keenness falls off. By comparison Widowhood will last from 10–12 weeks.

One or two fanciers have suggested that this Roundabout figure is the minimum and could be extended by a week or two. This does not actually matter because the Roundabout is used for the short races and then the birds are allowed to pair and lay so that by the end of the season they are going to the long races as Natural racers. Obviously you have to get the timing right which is where my table of 'Mating, Laying and Position for Racing' comes into its own. If you study the table you will see that if you want the birds to be sitting 10 days for the long race then their last on the Roundabout should be 3 weeks before. If they are easy come-back short races then you can risk the cock but I don't think it is a chance worth taking. Concentrate on building them both up for the long race. To me that is what is important.

What is also important about the Roundabout system is that for the small fancier it seems the best way forward. If you have the time and money for a large establishment you may be able to beat a small Roundabout loft, but not necessarily, because to do that you would need a stock loft, a sprint Widowhood team and a long-distance team on either the Widowhood or the Natural system, not to mention a young bird team and perhaps a yearling team. I once visited a famous German loft years ago where there were four full-time loft workers. The man owned department stores in Germany, so could afford it. I was talking to another fancier at the time, a French Count believe it or not, and I said to him at the time, adapting a famous French proverb, that it was magnificent but it was not pigeon racing!

EIGHT

The Off Season

The Moult – Fielding – The Open Loft – Separating the Sexes –
Weeding Out – Written Records – Reading – Repairs – Showing –
Putting Them Down – Show Teams – Show Racers – Pen Training –
Diseases – Canker – Coryza – Roup – Coccidiosis – Going Light –
PMV (Paramyxovirus) – Ornithosis, Chlamydiosis – Pigeon Fanciers'
Lung – Pigeon Pox – Injuries – Feather Parasites – Probiotics –
Electrolytes – Vitamins – Food Boosts

Most of this book up to now has been concerned with the months between January and September, the months when breeding and racing are taking place. There is an unfortunate tendency for some fanciers to think that this is the only important part of the year and that it does not really matter what happens during the autumn and winter. It is a mistake to neglect birds in these months, for no amount of work in the spring can undo the damage that has been done in the autumn.

The Moult

By September the birds are completing their moult, and the youngsters are going through this process for the first time. Good healthy pigeons should moult easily enough without any assistance, but if they are neglected or treated badly they can easily be set back so that their feathering will be less than perfect. If it is less than perfect in December it cannot possibly show an improvement before the first race. The birds, particularly late-breds, should have as much as they want of corn, green food, baths and exercise. They need not be forced to fly but a little flying on fine days will not hurt them. Some Belgians shut their birds up for weeks on end during the latter part of the moult, but this is not recommended. Lack of exercise will cause their muscles to get flabby, will upset their digestion, and their breathing will deteriorate. A moderate amount of exercise will not hurt them, but extended flying should be avoided.

Fielding

In the old days it was not uncommon for fanciers to open their loft in the autumn and let the birds go fielding. Birds were either not fed at home or only rarely in the evenings and they had to pick their food up in the fields. In this way it was thought that the birds were toughened up and a few pennies saved. One of the good old fanciers who used to practise this was Arthur Stow of Barrowford, but he lost his enthusiasm for it when his loft was virtually wiped out by the birds picking up poison. Arthur Stow raced in the days before seed dressings were as prevalent as they are today, and there are many fanciers who suffered during the craze for dieldrin. There are now some government controls on seed dressings but of course the risks are still greater today than they were in Arthur's day. This is only one of the dangers of allowing the birds to go fielding.

The greatest danger is that these birds may get shot. The law says that if a farmer sees your birds actually feeding on his crops, he can only shoot them when all other means of scaring them off have failed. That means that if a fancier turns his birds out field-ing, the law is on his side if a farmer shoots them, but I am afraid that I would be on the side of the farmer, for to my mind the practice of fielding does incalculable harm to the sport. Every time a farmer sees racing pigeons feeding from his fields, it is encouraging him to shoot them next time he sees them and this may be in a race. Even more seriously, it is encouraging him also to get the law altered so that he may have greater protection for his crops. Pigeon shooting is a serious menace to the sport, for champions get shot as easily as duffers and the practice of field-ing only makes the problems much more acute. There might be something more to be said for it if it really had beneficial effects, but very few if any successful fanciers practise fielding today. Its greatest use is that it enables some fanciers to keep birds during the winter at the farmers' expense. The very lack of success of this practice has caused it to be almost completely abandoned I am glad to say.

The Open Loft

Although fielding is now fortunately almost non-existent, many fanciers give their birds an open loft after and even during racing. The doors of the loft are opened in the morning and not shut until night. Birds are allowed to come and go and sit around on the roof for as long as they like. Sitting around on the top of the roof can be good for the birds, particularly when they can sit in the sunshine and absorb the health-giving rays. The birds like it and many fanciers think of it as giving the birds a treat after a season of hard work. The biggest danger is that if the birds are given an open loft in the autumn, they may learn to sit out and will not have unlearned the lesson when racing starts.

If the birds have been trained well to the corn tin, then they can be given a few weeks on the roof and will not suffer, providing at the end of each day they are made to come in smartly after their period outside. They can, for example, be given an open loft in the morning with the food hoppers down; the hoppers should be removed at late morning and not replaced until the birds are called in, about an hour before sunset. A system such as this can be worked quite successfully until September but, of course, as the evenings draw in so it becomes increasingly difficult. One of the important things to remember is that unless the birds have been very well trained they should not be called in on the corn tin unless they are hungry, otherwise the strict control which can be achieved by this method will be spoilt.

Separating the Sexes

The question of whether or not to give an open loft is complicated by whether the fancier has separated his birds or not. This is another matter which varies considerably from fancier to fancier. Some, particularly Natural racers, separate their birds immediately after those pairs have finished racing. Some, particularly Widowhood fanciers, keep them together longer to rear a pair of late-breds and separate them in August or September. Yet others do not separate them until January or February, shortly before they intend to re-mate them. Separation has two main ideas behind it. First, it is intended to stop breeding for that year, and second, it is intended to break up existing pairs so that each bird can be coupled with a different mate for the succeeding season.

If pigeons are left unseparated during the autumn then quite often they will stop breeding on their own account, but there will usually be a few birds driving and going to nest. For this reason most fanciers separate their birds as soon as the late-breds have been put in the young bird loft. Even while the late-breds are still in the nest, the unoccupied nestboxes should be shut up and this will discourage birds from going to nest. Separation has some disadvantages and the principal of these is that if the birds are to be given daily exercise, all the cocks must be back in the loft before the hens can be let out. If the open loft is permitted, they must spend alternate days inside.

Weeding Out

The time when the birds are being separated is, of course, the ideal time to go through a loft and weed out the birds that are not up to standard. If they are not good enough for you they will not be good enough for anyone else, so it is hardly worthwhile trying to sell them. It is at this time that a fancier must harden his heart and cull them. It may happen that when the fancier comes to separate them he may not have handled them for a week or two since he will have been busy with young birds, and in any case he should not have been handling birds in the moult more than is necessary. He should check them over to see that the moult is progressing favourably. He should look through their racing performances for this year and see whether they have earned a perch for the winter.

It does not matter what pure strain they belong to or how

much was paid for their grandparents, if they have not shown themselves worth their keep in racing, they must go. With some birds there may be some doubt, for the fancier may not have been able to afford to send a large team to the longer race points where entry fees are higher. These, if they have a pedigree of winning long-distance parents, are worth keeping. There is a great temptation to keep a bird whose parents were outstanding but which has not excelled itself in racing, for sometimes the good qualities in a pigeon seem to skip a generation. This bird may prove a useful breeder and this is the one possible exception to the rule, especially if it can be a possible mate for one of the birds intended to go the longest distances. Of course, this applies only to the son or daughter of a first-class pair and not to grandchildren or any other relationship. In the same way I would keep any bird that had flown 500 miles (800 km) in reasonable time more than once even though it might not have been outstanding in any race.

Written Records

For a fancier to be able to decide which birds to keep he should not rely on his memory. With a small team of birds a fancier will usually be able to remember the performances of every bird and will know by heart their breeding, but there is always a chance of a mistake, not so much with his best pair of birds but with those which are not quite so good but still very definitely worth keeping. Many fanciers are afraid of paperwork and keep little or no records and this is the cause of many mistakes. I have a sheet of paper ruled as a race sheet. On this I write down every bird, young and old, and put down not only every bird that is sent to a race, but every bird that is sent to a training toss. At a glance I can see exactly the amount of work that every pigeon has done that year. At the end of the year the sheet is used in selecting those to keep for the winter and is then put by for next year. A similar record is kept for young birds and the special value of this is that I can tell what tosses the late-breds have had and will not make a mistake of giving late-breds the same treatment as the earlier-bred birds.

I keep saying to myself that I am going to put it all on a computer and maybe one day I will. It makes a lot of sense. I remember many years ago the great F. W. S. (Frank) Hall gave me a son out of his champion K42. The bird bred me some good ones over the years and I still treasure the pedigree that Frank gave me. Written in his full round hand it was a calligrapher's work of art, but it proved to be a double-edged sword. As an auctioneer at his peak he was the best in the business but he was challenged by rivals who did not produce the same quality of advertisement. What a difference it would have been if computer programs had been available then. Call up the parents and the whole pedigree is there in seconds! The computer can include whatever you want (within reason): performances on the road and performances as a breeder. I am not going to advise on which program to get for the obvious reason that this is a rapid-growth area and new programs are coming out every few months. Any recommendation I make could be outmoded before this book even gets into print. Read the advertisements in *The Racing Pigeon*, or go to the big shows like the British National (Old Comrades) Show. Be prepared to spend a little time and do a little homework.

I know there are plenty of fanciers who will say that keeping these records is a lot of bother for something they know anyway, but in point of fact it is not all that amount of bother and the additional certainty it gives makes selection of pairs much easier and much more certain. In addition, two years later, when the memory is perhaps getting a little dim, I can check up on the work done for that year for any bird. On my racing sheets I mark the birds I pool and the order in which the birds return and this again is worth referring to in later years. One particular bird made its best performance from the 400-mile (640-km) race point; therefore the following year I gave it every encouragement and pooled it in the same race. Another bird was never well up in the shorter races, but in the longer races was always the first, second or third to the loft. Again we have that little bit more information and we are able to act accordingly.

While any information about position on the nest applies only to Natural birds, I believe that racing information about

Widowhood or Roundabout birds is just as important. Obviously position on the nest does not apply but little fragments of information, such as whether one needs to see his hen before he is basketed and for how long and whether this changes as the season progresses. One of the advantages of Widowhood for a fancier with a small loft is, it is claimed, that they can be treated as a flock not individuals. There may be some truth in this for sprinters but not so for long-distance Widowhood. In both cases I believe that while the sprint birds can be treated as a team, for distance birds it will pay to make a note of individual preferences and differences.

I go even further than that and I keep a record of the weather conditions on every race day and write a short report on the race, which I keep with the other papers. This, if it is kept accurately, is a good help in picking out those birds that are best on a hard day, those that are best in an east wind and those that are best in a west wind. Some fanciers race their pigeons like machines; all get the same treatment and the fancier does not distinguish one from the other any more than he does one maple pea from the next. I know several comparatively successful fanciers who do this. But I am sure that they could be even more successful if they considered their birds as individuals. What records do is help the memory separate the birds so that instead of being just a ring number, each pigeon becomes an individual with its own particular character. It is in knowing and recognizing each pigeon as an individual that half the pleasure of racing pigeons lies.

As well as performance records, breeding records should be kept. Most important of these is a loft book. As I keep only a small team, I can in fact use the loft book pages in *Squills Year Book.* Those with large lofts will have to buy a special book. I would buy the cheapest because it will be battered by everyday use and at the end of the year you will be transferring it to your permanent record books or computer. When the birds are mated up, the ring numbers of the sire and dam are entered and as the youngsters are hatched and eventually ringed, so the ring numbers of the progeny are entered. There are spaces for the date of laying and the date of hatching which can provide valuable information if the old birds are to be brought into some special condition for racing. In addition, of course, the information of the ring numbers is the basis of the pedigrees. Pedigrees are nearly always required by a fancier buying pigeons, but apart from this they are well worth keeping since by studying them it is quite often possible to make a good mating which would not be immediately apparent by just handling the birds in the loft. Sometimes it is easy to overestimate the importance of the pedigrees. More often than not they are little more than a waste of paper, but if the breeding has followed a certain plan and if the fancier really knows what he is about, they can be a help. Nevertheless it should be always borne in mind that pedigree is never more important than performance.

Reading

The winter months are a good time for a fancier to make sure that all his records are up to date. It is also a good time for him to catch up on his reading. Unless a fancier is right at the top of the tree, he cannot afford to ignore books and magazines published on pigeon racing and in the winter he should read some of the books he has not yet read and some of the best books again. That does not mean, of course, that he should believe everything he reads, because many pigeon books and most articles by successful fanciers give only the methods which have brought them success with their own birds. They will not necessarily give success to another fancier with different birds. Nowadays there are a good number of videos available and many of them are worth looking at. The trouble is that at £15 to £20 they are relatively more expensive than books and they are not all equally good. If a new book comes out you can go to one of the big shows and flick through the pages to find out if it is right for you. I have looked at many videos and some of the visits to the lofts of successful fanciers are informative but not all tell you what you want to know. If, like me, you are a fan of Geoff Kirkland, who I think is one of the best minds in pigeon racing today, you will want to see his videos. The first one, which he shares with Frank Tasker, another original mind incidentally, is very good and in fact much more informative than its sequel. This is only a personal opinion but I am talking about a loft I have visited and photographed. I don't know what to advise other than to go to big shows and see if the producers of the video are there and will let you see the one you are interested in. Other than that ask other members of your own club; you may be able to borrow one and if it's any good buy it because the good ones need several, indeed many, viewings to get the best out of them.

Nevertheless if a fancier reads intelligently, separating the wheat from the chaff, he may be able to find that something which will enable him to improve his performances. That does not mean to say that as soon as he reads one winning method he should immediately go and practise that method; then reading of another method, immediately change over. To act like this is just asking for trouble. A fancier must read and then think before he acts.

Another winter job for the fancier is to plan on paper next year's matings. Pairs can be mated and unmated on paper, pedigrees and performances of prospective sires and dams can be compared and the programme for the following year prepared. In this way, without making mistakes because of haste, next year's pairs can be decided well in advance of the busy period.

Repairs

At the same time, the autumn and winter weekends, when they are fine enough, are the ideal time to do repairs and alterations to the loft. A coat of white in the loft when the birds are separated, a coat of paint outside, wires and dowels replaced, the roof made watertight and all the odd little jobs that never get done in the season should be done well before racing starts. Pigeons can be very funny when they are coming back from the races; a new trap, a new strip of felt on the roof can all upset them and lose valuable seconds during a race. If these jobs are done during the autumn, then by the time racing starts, the birds are accustomed to their changed surroundings.

Showing

Showing is an occupation that can give many hours of pleasure to a fancier, but it is something that must be gone into carefully if there is to be any hope of winning. If all a fancier wants to do is to take a few birds along to the local club in the winter, have a chat with the members of the club and then go home again, not really worrying whether he has won or not, then no special trouble need be taken. If, however, he wishes to enter for one of the great national shows, like the British National (Old Comrades) Show, and win there, then he must go about it in the proper way. He must have the right pigeons and they must be put down properly.

Putting Them Down

Far too few would-be exhibitors realize the importance of putting birds down properly, and even at the best shows in a class of fifty or sixty birds, quite often there will not be more than half which are worth considering on this account. First of all, they must be clean and free from ticks or lice. This is a simple enough matter but one which is sometimes unforgivably neglected. Second, they must not have any broken flights. It is possible that a bird kept in a wire-netting cage may break its flight or bend it. If it is bent, then the wing tip should be put in the steam from the spout of a kettle, so that it straightens itself.

Fret marks may disqualify a bird but this will depend on the judge and the quality of the birds in the class. If a bird's only blemish is a fret mark, then it should certainly be entered in a raced class. Pin-holes in the tail are another blemish that will separate birds and decide the winner. They are caused by a parasite which can be cured, although the damage will have been done for that year. The general condition of the plumage is important as well. A bird which has been out in the misty wet towns, racing in all kinds of weather, cannot hope to have the same fine condition of plumage as a bird that has only been allowed to fly in good weather. This brings us to the crux of showing, the fact that it is difficult, if not impossible, to race and show the same birds in the same year. I am talking now, of course, about showing in the national shows. It is possible, but the fancier is making it difficult for himself.

Show Teams

If a fancier really wants to go in for showing, he would be well advised to get a separate team, or just a pair of birds which he keeps for the shows. John Thornton, one of Scotland's most successful showmen a few years ago, raced and showed the same family of pigeons. There was in fact only a slight difference between his racing birds and his show birds. The youngsters were raised and those that looked as if they would be most successful in the pen were separated into lofts and aviaries set aside from the race birds. From then on these show birds received different treatment from the ones with which he continued to race.

There are, of course, no hard and fast rules. A bird may be changed from one loft back into the other according to whether or not it looks most likely to gain honour on the road or in the pen, but they are not usually changed until the end of the season. The same thing can be done on a smaller scale by even a novice fancier with a small number of birds, if he is prepared to go to the trouble of having a separate system of management for perhaps only one pair of birds. It is the only way that he can hope to compete in the best competition.

Show Racers

The danger of showing birds is that a fancier may be attracted too much to appearance when he is mating up his racers and he may begin to believe that he can produce show racers equally successful in the pen and on the road at the same time. Without going so far as to say this is impossible, I would say that it certainly is, as far as the ordinary fancier is concerned, very, very difficult. For a racing bird there must be one criterion only and that is performance. Its breeding ability and its appearance must not be considered. For a breeding bird, that is a stock bird kept to produce only, there is again one criterion and that is the success of its progeny. Its pedigree, its own success on the road and its appearance must be ignored.

In the third class of birds, the show birds, the appearance and handling is the one standard by which it must be judged. If the fancier diverts for a moment from these standards he is heading almost certainly for mediocrity, and, if he has once held pride of place at the top of the tree, will soon find himself falling. The same warning must be given to those who decide to keep a few of the rarer-colour birds, the yellows, the silvers and the duns. If these birds are to be used in racing, then they must be judged on their racing merits and the colour ignored. If they are to be kept mostly for the pleasure of their colouring, then the fancier should not worry too much if, when racing, their performance is not all that he might wish.

Pen Training

There is one more point of considerable importance, and that is the necessity of seeing the birds have adequate training in the pen. This means they must be accustomed to being closed up in a pen near other birds and should accept it without anxiety, resting quietly. A bird that, as soon as it is penned, flutters madly will soon destroy all its hopes of winning. The fancier should buy himself a set of pens and should give his birds an hour or so inside them for a number of days, until they will rest quietly. Another important precaution is to train the birds to be quiet in the hand. Show birds should be handled often by the fancier so that when they are picked up they rest quietly in the hands without struggling. To do this, a few of our top fanciers allow even their best birds to be handled by fanciers who visit them, and after the main shows allow them to be handled even by novices, who at times can be less skilled, less gentle. It is good training for the birds, and it helps to keep them winning in the best of competition.

If showing appeals to you then you should try and get hold of one of Doug McClary's books. He is a good writer and his two books should set you on the right path. Unfortunately they are both out of print but your local library may be able to find a copy of *The Show Racer*.

Diseases

Pigeons, like any other living thing, may become diseased in one way or another and there are two ways popularly used to combat disease. The first and probably the most effective is to remove and kill all diseased birds immediately they are suspected. In this way the disease is immediately checked from spreading and no time is wasted on cures. It is argued that the bird must have been weaker than its fellows to have been stricken by the disease, and it would not be worth keeping in any case. These arguments are valid up to a point, but only up to a point.

A bird is not necessarily weaker and a poorer racer because it has been stricken by disease. To make a comparison with ordinary people, young birds, like young children, can frequently have disease and continue with perfect health for the remainder of their lives. In many cases a fancier cannot afford to keep a bird until it is fit again, particularly if its weakness has prevented it being raced that year. A year's feeding for a doubtful bird is an expensive luxury. The alternative to culling is curing and in many cases this is increasingly possible. Even those who proclaim most strongly that it is no use trying to cure sick pigeons cannot hold their ground when an epidemic spreads through the loft as it may easily do.

Pigeons can suffer from a wide variety of diseases. About some of these a lot is known and there are some about which there is only guesswork. In recent years the situation has improved enormously and paradoxically a lot of it is due to the Ministry of Agriculture's failure to understand pigeons and inability or unwillingness to invest in research. The Ministry has an understandable duty to protect the poultry industry in this country. It therefore becomes paranoid at certain things that it thinks may be detrimental to that poultry industry. We have seen it in the past with Fowl Pest regulations which have been imposed on pigeons without any attempt to evaluate the real as distinct from the theoretical risks.

With the arrival of PMV (paramyxovirus) in this country the Ministry went to town. I went to meetings at the Ministry in the early days and attempts were made to convince me that the outbreak of this worldwide epidemic was from the pigeon fanciers in the Sudan! When I tried to explain that there were no pigeon fanciers in the Sudan the officials did not want to know. When I went further and said that some fully qualified vets in Egypt who were also pigeon fanciers had seen the disease in the Middle East before the first Ministry reports they still didn't want to know. Unlike in the Sudan, there were small numbers of racing pigeon fanciers in Saudi Arabia and the Lebanon, as well as Egypt, and of course since historical times there were also semi-domesticated pigeons. I wasted my time. The Ministry's mind was made up and they did not want it confused with facts.

I shall be writing about PMV later but the outcome of the Ministry's meanderings was that a PMV vaccine industry was created. And because the Ministry was floundering between the two stools of prescription-only treatments and the practical supervision of thousands of fanciers, all of a sudden there were fully qualified vets being paid to interest themselves in pigeons. There had always been a handful of vets who were pigeon fanciers and these pioneers deserve full credit. One of these was Dr Leon Whitney, whose book, now outdated, was a classic of its time. The earlier books, including my grandfather's, are just not worth reading nowadays. With the increased 'pigeon awareness' among vets in this country came the first edition of the brilliant Dr Ludvig Schrag's book *Healthy Pigeons*, which included full-colour illustrations and was printed in many languages. Regrettably, Dr Schrag died in 1994 and the sixth edition is the most recent, but his book set a new standard in handbooks on pigeon diseases. There have been others, but his was the one that shaped the future.

The result of these influences has been no less than a revolution in the management of pigeon disease. Before the revolution there were few vets capable of helping pigeon fanciers and few fanciers willing to pay for frequently uninformed advice. After the revolution there was an increased number of vets who knew about pigeons and a greater willingness on the part of fanciers to pay proper fees. There must be now hundreds, indeed thousands, of fanciers who regularly consult their vet. Best of all, many of the pharmaceutical companies have set up free telephone help lines, and for the cost of a telephone call you can get advice on any problem in your loft. They are not just pushing their own products either. I once had a problem in my loft because while I was away on holiday the birds were not cleaned out. When I came back there was a hairy fungus, *Aspergillus*, all over the droppings. I asked the company Harkers what I should do, because not only did it look alarming but it could have been passed on to humans, i.e. me. The suggestion was iodine, the ordinary tincture available from chemists for pennies rather than pounds. The dose was 1 ml (measured by a hypodermic syringe) in the usual size drinker. It worked like magic. I wish I had asked years ago, but the point of the story is that this information was available to me, or anyone else, free.

Traditionally the principal diseases liable to attack the birds were canker, roup, coccidiosis and one-eyed cold. These are the names that are given to them by fanciers, but what exactly each fancier means by canker and roup will be found to vary.

Canker

On canker there has been a considerable amount of research and although canker can occur through other causes, the organism *Trichomonas gallinae* (or TG for short) is one of the principal causes. The disease is most obviously visible as a yellow or white scabby deposit in the mouth and throat or as a lump at the vent. The deposits on the mouth build up if unchecked so as eventually to block the throat. The organism TG can also cause internal infection particularly to the liver although almost all the organs can be damaged, even without any noticeable change in the throat of the bird. Only the lower digestive organs are not affected and this distinguishes it from coccidiosis, which is detectable here. A cure for this form of canker is a commercial product containing amino-nitrothiozole. The disease is passed from bird to bird through the drinking water and therefore either the other birds must be treated or the infected bird must be separated. The treatment has the advantage of immunizing the bird from further attacks. The good news is that since a cure for canker has been available for a number of years, the disease seems to be passing into the history books. I have not had a case in my own lofts for ten years or more, and I think most of my clubmates would say the same.

Coryza

Both roup and one-eyed cold belong to the group of bird diseases known as coryza. Similar diseases can usually be found in birds suffering from colds, sneezes and greasy wattle. The last, of course, is the discolouration of the wattle owing to ill-health and should not be confused with the dirtiness of the wattle that occurs while the bird is feeding youngsters. A simple cold may cause the bird to have its nostrils blocked with mucus. The pigeon can be made to 'blow its nose' by tickling the nostrils with a feather. If a primary which has been moulted out is pushed fairly hard up the nostril, the bird will sneeze and clear its nose. Different feathers should be used for each bird so that infection is not spread.

As with canker so with coryza there has been a change and nowadays it is likely to be referred to as NSRD, Non-Specific Respiratory Disease. This is a huge area into which I don't intend to venture. The words Non-Specific really mean that the vets haven't a clue about what causes it. They will say that there are many organisms responsible but that they cannot isolate which one in any specific case. The answer is to give a broad spectrum treatment. All this means is that you give a medicine that works for a number of related illnesses on the assumption that your problem will be within that number. It works and so far there have been no bad side-effects so it has got to be worth trying. There are commercial treatments advertised and others, like

Tylan, that have been found to work but are available only on prescription from a vet.

Roup

Roup is in some respects similar in appearance to canker, particularly because there may be deposits in the throat, but in roup these are accompanied by mucus in the nostrils. As is the case with canker, the deposits if large enough can be removed by a matchstick or tweezers and, of course, the bird should be separated. Perhaps the most infectious of the coryza group is one-eyed cold. This shows itself as a badly watering eye and by inflammation of the eyelids. Almost invariably this takes place only in one eye, hence the name, although sometimes both eyes can be affected. For easing the eye, bathing with a boracic acid powder solution is usually satisfactory, and if the birds are given half an aspirin then this will assist recovery. Soluble aspirin is usually the best form to give. There are several commercial preparations for dealing with all coryza infections and those based on a preparation of furazolidone seem most effective.

Nowadays it's only old fanciers who talk about roup. The younger ones prefer some pseudo-scientific name that disguises the fact that we don't know a lot about it. I prefer the old name for another reason. The pen name Squills as in *Squills Year Book* and Squills Food for Novices in the weekly *Racing Pigeon* was used first by my grandfather in the 1870s when he was writing in *The Fancier's Gazette*. It was a joke because he was looking for a cure for roup and someone misheard him and gave him a herbal cure for the old children's illness croup, also a respiratory disease. Grandfather gave this herbal remedy, Syrup of Squills, as a cure for roup not croup, much to the amusement of his friends, and the name stuck. Funnily enough, the joke is now on them because this self-same Syrup of Squills is today being prescribed by herbalists for respiratory problems. Just to show I'm really sentimental at heart, near my loft is a patch of snowdrops and the pretty blue flowers of *Scilla siberica*, the very plant used to make Syrup of Squills.

Coccidiosis

Coccidiosis is another disease which can cause needless trouble. It is usually present in a mild form but rarely is it fatal. Coccidiosis is normally the result of unclean lofts, although it is not unknown in the cleanest and best-organized lofts. It is almost always the result of birds picking up food which has been soiled by droppings. The cure for this, apart from making sure that the loft is spotlessly clean, is to put in the birds' drinking water a solution of sulphamezathine or its more modern versions, or one of the other antibiotics available on prescription. Nowadays the detection of the oocytes (eggs) of coccidiosis is easy with a microscope and its cure is not difficult. In the old days many fanciers also regularly gave their birds a diluted drink of permanganate of potash once a week. In Belgium it is often kept in a concentrated form in an old wine bottle and is therefore known as *Vin Rouge*, red wine. Like so many of the old cures, this has gone out of fashion although permanganate of potash was a good sterilizer. I gave it up because it stained the drinkers and they never looked clean. Its place has been taken by any number of water sterilizers, all more expensive but I wonder if they are more effective. When I get to the paragraphs on probiotics, I will give another reason for being very careful about the use of sterilizers in the drinking water.

Going Light

Another disease that is extremely troublesome and, if possible, even more difficult to diagnose is going light. In many cases the bird loses weight because of failure to eat and digest. The bird continues to lose weight even though it has food in its crop. This seems to indicate some sort of digestive ailment, and a dose of Epsom Salts will often put the bird on the road to recovery.

Another cause of going light is worms. These parasites gather in the intestines and live on all the food that should be going to the bird. Many poultry worm cures will clear these and there are now medicines specially prepared for pigeons. These come in two forms, capsules or liquid, and I am afraid that the idea of giving forty capsules to forty pigeons is not my idea of fun and I only have about forty birds. I much prefer to tip the medicine in the water if it is necessary. If a vet diagnosed a serious infestation in an important bird then I might use capsules. If I have a bad race with a number of birds out overnight, then I might be tempted to use the liquid form, but I wish I knew how bad the problem was. Many years ago I had a friend who was training to be a vet and as her special subject she chose worms in street pigeons. She caught the pigeons, killed them and dissected them in my old loft in Doughty Street. I was amazed and sickened as pigeon after pigeon was cut open to reveal a seething mass of worms. I realized then why you never saw much wattle on a streeter; they all died of worms as youngsters and yearlings. A few racers were sacrificed for comparison and not one of these showed worms.

One worming a year before the breeding season is a precaution that some like to take but it must be accompanied by strict hygiene. Once a pigeon has worms they will multiply inside the bird until the bird dies. Parts of the worm break off and can be easily detected in the droppings using a microscope. The infection spreads by worm eggs, that contaminate the food. If you have a massive infection then the loft floor must be disinfected. The old-fashioned rubber glove treatment was to scrub the floor with a boiling-hot solution of caustic soda but there are more modern and less drastic disinfectants suitable. What is absolutely vital is to prevent the food getting contaminated.

In the winter when the fancier goes to and returns from work in the dark, many have little choice but to fill the hoppers in darkness. If it is a mixture it can get thrown on to the floor as the birds sort the feed out to reach their favourite grain. Some goes on the floor when it may come into contact with the eggs of worms or the eggs of coccidiosis come to that. Ordinary scraping out even if daily, which is unlikely in winter, will not get rid of the disease eggs. I have tried giving the birds single grains such as beans only or barley only but the feed is still thrown on the ground. My solution is far from perfect but seems to work. I cut the food right down. As simple as that. Then birds are so hungry they eat the food quickly without scattering it and if any should get thrown to the floor it is picked up before it has had time to get contaminated. When I was not able to feed them in daylight in winter, it seemed to work. Even now when I can feed in winter daylight I still use the same idea. However, because 100 per cent freedom from worms cannot be guaranteed, I treat the birds with a de-wormer before they go to nest and take particular care to make sure there is no contamination.

Yet another cause of going light is avian tuberculosis. This usually occurs in adult birds although youngsters may be affected. Post-mortem examination will show characteristic tubercles on organs such as the liver and lungs. Birds suffering from tuberculosis must be separated at once. They cannot be properly cured and must be killed before they spread infection. Tuberculosis can attack birds of all ages, but they go light slowly. Where youngsters go light rapidly, paratyphoid can be suspected, particularly if the old birds are suffering from swollen joints. In these cases, all birds should be treated, as birds can be infected without visible symptoms. Eggs can also be infected, surprisingly enough. Until recently there was no cure for paratyphoid, but it is probable that sulphathiozole or another of the sulpha drugs would cure it, since these are effective in some similar salmonella diseases of poultry.

The swollen wing joints of paratyphoid cause more arguments than any other disease. A fancier buys a couple of pigeons from another breeder. The pigeon looks healthy, but within weeks these new birds have developed swelling in the wing joints although all the original birds in both lofts are unaffected. The fancier thinks that he has been sold a pair of diseased birds and is naturally upset but in fact the purchaser's loft was the one at fault. The disease paratyphoid, like many other salmonella diseases, is extremely widespread, but the birds can immunize themselves against the disease. They are exposed to the disease in such a mild form that it is unnoticed in the loft and as a result they build up an immunity just as effectively as if a hypodermic had been used. As a result the disease is running throughout the loft, but because the birds have immunized themselves there is no visible sign of disease. Enter the two birds from another loft where there has been no disease and down they go, with disastrous consequences. This theory is by no means accepted by the veterinary profession but a better one has yet to be put forward. If my theory is correct there is not much point in any mass curative programme. The great encyclopaedist Wendell Levi had his own cure. He said that if the main wing flights were pulled then the bird would cure itself. I have never tried it but I would be interested to hear from anyone who has because Wendell Levi was a very wise fancier.

PMV (Paramyxovirus)

If you have been living on Mars for a few years you would not know about PMV. I have already voiced my disquiet about its history, now to the practicalities. Within the RPRA (Royal Pigeon Racing Association) and most other European pigeon organizations, vaccination against PMV is compulsory. The regulations in each country vary so I will limit myself to the RPRA procedures that I have to follow. The rules at the time of writing prohibit the use of oral vaccines although they are permitted in some parts of Europe. These vaccines are simply put in the drinking water but like all drinking-water treatments there is no guarantee that every bird will have taken the right dose. It is also claimed that an oral vaccine only works for a matter of weeks or months before it must be repeated. All the vaccines accepted by the Ministry (MAFF) are injected via a hypodermic. These vaccines fall into two groups: those that are oil based and those that are water based. Most fanciers prefer the water based because it is easier to inject. It is also claimed that it does not soil the feathers, as oil-based ones do, but I think this is advertisers' hype rather than a real benefit. There used to be one vaccine that had to be injected in two parts, separated by several days but fortunately all the modern vaccines are one-shot, and it seems to be recognized that once a year is enough. The RPRA requires only that race birds be vaccinated so that stock birds can be left alone but most fanciers vaccinate the whole loft in the winter. Young birds must be vaccinated before racing so many fanciers do this about the time of the early training tosses. In short PMV is no longer a problem since vaccination, for better or worse, has become a way of life.

Ornithosis, Chlamydiosis

Ornithosis simply means bird disease; psittacosis means parrot disease. The best term is chlamydiosis, which means caused by the organism *Chlamydia*. Parrot disease has a long and serious history going back to the smuggling of parrots and budgerigars out of South America. Because it is an emotive disease likely to catch the eye of planning and health officials, I have tried to check whether there are any real risks to pigeon fanciers. It is not helped by newspaper headlines like 'Father of five killed by pigeons next door'. This Somerset case caused me hours of work but in the end I got an apology from the coroner, the chief constable and the newspaper because there was not a word of truth in the headline. Similarly and just as seriously the headline 'Railway guards at risk from pigeon in guards van' appeared in one of the medical journals. This is going back into history thirty or forty years, when we still trained on the railways, but no matter how old, these bad stories will live for ever in the medical literature if not stopped. In this case two brothers living in the same house had suffered from ornithosis so although the plural guards is technically correct it was only one possible infection. No pigeons were ever examined and, unbelievably, the two brothers owned a smuggled parrot that had died at the time from an unknown disease! Years of study have led me to believe that ornithosis is not a real problem for pigeon fanciers.

Another reason for my interest is that ever since I can remember I have had some sort of chest and nose problem. My father had it too and it was then called post-nasal drip. My two brothers never had it so if it was hereditary I drew the short straw. As a result, I have been in and out of clinics all my life, the first of which was the North Middlesex Hospital Chest Clinic in the fifties where they did the old-fashioned scratch tests to find out what I was allergic to. There was no result with pigeons but there was a faint reaction to *Aspergillus* as I have already mentioned. It was here that I started to get seriously interested in the general health implications for all fanciers. The doctor told me the pigeon test was made up from pigeon feathers ground up in a sort of medical Moulinex. I suggested the problem might come from bloom and wondered if body feathers with a higher proportion of bloom might have given a different result.

Years later we were approached by Dr Pepys of the Brompton Chest Hospital in London and at one of The Old Comrades Shows held then at Alexandra Palace we persuaded hundreds of fanciers to give a blood sample. No ornithosis showed but as a result *The Racing Pigeon* set up a committee with fanciers and doctors to discuss respiratory problems. At about the same time in Glasgow the Scottish National FC was working with Glasgow University and research was going on in the Plymouth Hospital. Of these projects only that in Glasgow is still going and research is centred on pigeon fanciers' lung, so although this is not a disease of pigeons, it deserves a place here now.

Pigeon Fanciers' Lung

This is one of a series of dust-related diseases of humans. There is farmers' lung and dozens of others, with the best known being miners' lung or silicosis. The effects of miners' lung are familiar to all older people and to all who read social histories or even fiction about miners. Very simply the miners breathed in coal dust which permanently damaged their lungs. Since there are now so few mines and so much more awareness, the number one lung disease is now asbestosis.

All these diseases are treatable but not curable, because the damage to the lungs is irreversible. Prevention and early detection are therefore vital. This is why face masks are being used more and more in pigeon lofts. The modern trend of closed-in lofts has meant that there is greater risk of dust and it makes sense to use a mask when scraping out even if it is not used during normal loft visits. There was an old practice of putting sand and lime on the floor and then putting it through a riddle or sieve every week to get the droppings out. Even with a mask on this is looking for trouble.

What I have written may sound worrying but very few fanciers have any trouble and nowadays doctors realize they cannot just say give up your hobby. One of the key tests is the annual holiday. You go away for two weeks hopefully in the sun away from all pigeons and then come back to the reality of cleaning out. If when you go into the loft you feel wheezy, sweaty or unwell in any way that is the time to be tested. Your local GP can take a blood sample and send it to Glasgow for examination. The address is Research Co-ordinator, Research Office, Room 48, Dept of Respiratory Medicine, Stobhill NHS Trust, Glasgow G21 3UW (phone 0141 201 3724 or 0141 558 1218). You can also, as I do, have regular tests. At the Blackpool Show and at others the Glasgow team led by Dr Lynch and Dr Boyd will be there painlessly taking samples free. They also have pamphlets available by post or at the shows. Let me make it clear – pigeon fanciers' lung is not a simple allergy for an allergy need have no permanent effects. PFL on the other hand can inflict permanent damage. Wear that mask!

Pigeon Pox

This can be a problem disease because there is no known cure for the pox virus. The typical pox scabs on the non-feathered parts of the birds and the bacterial secondary infection can be treated by one of the chlorotetracyclin group of drugs obtainable from the vet. The good news, or at least the better news, is that the disease mostly attacks young birds and that if the bird survives one attack it has a lifetime immunity. The virus form seems to be highly infectious and every year one hears of an outbreak in a local club or Federation during the racing season. This is not surprising considering the unhygienic way we transport birds. If there is an epidemic like this it would be impractical to treat the whole of a club or Federation. They should stop racing and wait until the epidemic has passed. This means the whole of a club or Federation because almost certainly those birds that do not look infected are carrying the disease. It may be necessary to kill those very badly affected or go to the vet for the antibiotics.

My experience is that most birds are only mildly ill and that they will get over it. Unfortunately I don't know how long it will take to run its course. It is going to be weeks so probably the best advice is to shut up shop for the rest of the young bird season. I keep hearing stories of fanciers who keep the obviously infected birds at home but race those that look all right. This is wrong and downright irresponsible because even those birds that look well are carriers which could infect everybody else's birds.

Pigeon pox can be prevented by vaccination; you pluck out a few feathers from the bird and paint the vaccine on. If your neighbour has pox in his loft and you are as yet unaffected, this may be worth doing. You should know if the vaccine has taken in one week and should be able to basket them in two weeks at least for training tosses. The snag is that vaccination lasts for only one year whereas auto-immunity from infection and recovery lasts for a lifetime. Like all these treatments the birds need an R and R programme (Rest and Recovery) as I explain in a later section.

Injuries

Injuries in a bird are much more easily healed than would seem possible. A bird can hit a wire and gash itself badly but recover in next to no time. This is partly because of the high body temperature and more particularly because of the fast blood circulation inside the bird. A wound can be assisted to heal by stitching. An ordinary curved surgical needle and surgical catgut should be used. Injuries are also caused by birds pecking at the heads and eyes of others, particularly in squabbles concerning the possession of a nestbox and these will generally heal successfully if bathed in a boracic acid solution and further fighting is prevented.

A broken leg can occur if the bird alights on wire netting or if the bird's leg gets caught in a faulty basket. For this reason medium-small mesh netting should be avoided in a loft, and all baskets should be lined with sacking by the manufacturer. A broken leg can be set by the fancier. Splints are made from matchsticks or small pieces of wood and the leg bound up with sticking plaster. Some skill is needed in getting the broken bone properly lined up, and for this reason I prefer to leave this job to those with more experience.

In some of the bigger towns there are clinics run by the PDSA (People's Dispensary for Sick Animals) and RSPCA (Royal Society for the Prevention of Cruelty to Animals) where this work is undertaken. The PDSA is by far the best bet because the RSPCA is not particularly pigeon minded. (You try and get the RSPCA to help you if a cat is terrorizing your pigeons. Not a hope and the cynical might say it had some connection with the fact that cat lovers give more donations than pigeon fanciers.)

If you do treat the birds yourself, the bird should be given half an aspirin and a box of its own so that it may rest. Old birds sometimes can cause false alarms as it is not uncommon for them to be affected with some sort of rheumatism in the leg which makes them limp, but if they are rested and fed well for a week or so, this will soon pass. I have seen advertised a sort of plastic harness for suspending the bird and keeping the weight off the leg and I am told it works well. I was talking about this to an old fancier and he said to try cutting holes in my wife's tights! If the bird is important do not hesitate to go to the vet fast because the bone can set badly very quickly.

Feather Parasites

The disease which is likely to cause havoc if it arrives in the loft is feather rot. This is caused by a parasite that lives on the bird. There are many different varieties of these, including some that eat the feathers. Some fanciers declare it cannot be cured, but in point of fact it quite frequently yields to a treatment by a sodium sulphide or potassium sulphide (NB not sulphite) solution. This is made of 1 gallon (4.5 l) of tepid water with 0.75–1 oz (21–28 g) of chemically pure sodium sulphide and a small piece of soap or a few drops of detergent as a wetting agent. The bird is dipped in this solution for 20–30 seconds, leaving out only the head, and thoroughly ruffling the feathers. Then the bird's head is ducked once or twice, keeping it in the solution only for an instant each time. This must be done carefully, since the solution is a poison.

There are many different types of lice or ticks which will live on birds if they are not kept in check. Young birds in the nest are liable to be attacked and can usually be cured by a dusting of insecticide powder. A natural powder like pyrethrum is much to be preferred here. If a pigeon is found to be suffering from ticks, it should be dusted, the nestbowl should be changed, the fresh sawdust powdered and the insecticide blown into all the cracks and crevices of the nestbox.

The smaller ticks include red mite, tiny creatures which live on the blood of the birds. They do not carry disease as some of the others do. They are, however, difficult to clear from the loft since under laboratory conditions they have lived 20 months without food! In the past, red mite were difficult to eradicate but liquids, like malathion, sprayed once a year can usually keep these under control. Malathion can be extremely dangerous if not used properly. I once heard of a convoyer who put it in his bath water and I fear that his doctor would not have approved of this. DDT is now banned because it tends to remain in the soil and is ecologically harmful. A lot of the ordinary powders are now based on pyrethrum which acts very rapidly and does less long-term damage to the environment.

A common mistake made in the treatment of all diseases and pests is the belief in washing the loft with antiseptic solution as a cure-all or prevent-all. This as a rule does not have the effect that is hoped for. Many diseases spread in damp conditions and the washing down of the loft actually helps to carry them from one part of the loft to another. Although the antiseptics are effective against certain germs, they by no means eliminate all undesirable organisms.

Probiotics

Most people know what antibiotics are. To put it crudely they are chemicals, drugs to kill disease germs. Unfortunately they can kill off some of the good ones as well as the bad ones. The different proprietary brands of probiotics have different formulas, but basically they consist of dried bacteria, the good ones, that can be brought back to life by the natural moisture inside the pigeon. The bacteria chosen are those that form the intestinal flora of the birds. They are the ones that are necessary for the proper digestion of all the nutrients in the food. When you administer antibiotics, you also destroy the flora; the probiotics replace them. Most of the time they are not needed but after medication and after severe stress they will speed recovery. Severe stress in this case includes hard races, those where the bird spends a night out and, in general, anything a pigeon might find upsetting. They should be given usually the day after medication, vaccination or the hard race but the different products vary so read the packet. Some of the probiotics also include vitamins so again read the packet because if you are going to give a vitamin supplement you may want to choose the day in the week you give it according to when you gave the probiotic.

Electrolytes

Call me old fashioned but I don't like all these new names. For years we talked about minerals. Now exactly the same things are called electrolytes, trace elements, or sometimes ionites. These minerals or mineral salts cover a wide range, including sodium, potassium, magnesium, calcium, citrate, sulphates and lactate. All of these are found in the pigeons' natural food and calcium, for example, is an essential component of grit. There should be no need to feed them on a daily basis but there is a good case for having them in reserve to use after a bird has been under stress either from medication or a hard race. I am not a scientist and I am not sure there is any scientist in the world who could say exactly what minerals are needed and in what quantity. We do need minerals and since the amounts obtained from natural sources may vary there is a case for the occasional use of mineral supplements or electrolytes. Many of them are combined with vitamin supplements which makes them even harder to evaluate.

Vitamins

In the section on food I wrote about vitamins as they occurred naturally, so this section is about vitamin supplements. These can come on their own or mixed with probiotics, electrolytes or even food supplements. To recap, vitamins B and C are vital to a pigeon's health but being water soluble are not retained in the body. Also since they are water soluble and are not retained in the body there is no risk of overdosing and if too much is given it just goes out in the droppings. Harmful only to the pocket.

The other vitamins are oil or fat based and will be retained in the body for a while, so need not be administered in the same way as the water-soluble ones on a regular basis. It is possible to overdose in theory but in fact I have never heard of it. During the racing season, particularly with long or hard races, a supplement is advised. It is also useful during the breeding season when obviously there are exceptional demands. I wish I could advise more positively about the use of supplements. The manufacturers are going to want you to use them to the maximum. I am looking for the minimum. As a rule of thumb I would say: Never more than once a week; Always after heavy stress.

Food Boosts

To conclude this particular section I should mention the comparatively recent development of food boosters. The idea of carbohydrate loading is as old as pigeon racing. In ancient times a fancier simply gave his birds more maize just before the race and would have been amazed if you had told him he was carbohydrate loading. In recent times the use of weak mixtures and strong mixtures includes a strong element of carbohydrate loading from maize as well as an increase in the protein. The boosts I am thinking of are the concentrates over and above feeding changes. Vydex Carbosol is one such, a soluble carbohydrate liquid with a high concentration of easily available energy supplies. This is not the same as the old trick of putting glucose in the water since, according to their research, glucose increases insulin levels and hinders the absorption of the glucose sugar energy.

NINE

Officials and their Work

*Club Organization – Members' Duties – Federations and
Combines – The Unions – Courts of Appeal – Leading Cases – Velocity
Calculation – Corrected Times – Summary of Method – An
Example – Approximate Velocities – Decimal Places – The Accuracy
of Distances – Dead Heats*

Although clubs and federations have been talked about before, in this chapter the organization and administration of the sport in Great Britain will be considered. The most important part of this organization is the club, and its principal officers are four: the Secretary, the President, the Chairman and the Treasurer. In many clubs, the President and Chairman are one and the same person and so are the Secretary and the Treasurer. There may be other officials elected, such as press secretary, ring secretary, race secretary and show secretary, but this is by no means normal. An assistant secretary is also fairly frequently elected and is a very useful person, since on the whole the secretary has more than enough to handle.

Club Organization

A few secretaries are unpaid, but nowadays most receive an honorarium usually based on the membership of the club which is too often very little. In spite of this, secretaries are still not easy to come by and good ones are invaluable to the success of the club. Although a good secretary is often given a fairly free rein, always the controlling power in a club should be the committee. This may consist in a larger club of a number of elected members but quite frequently in smaller clubs consists of all members of the club present. This is a convenient arrangement since it allows decisions to be reached rapidly.

A very important factor in the day-to-day work of the club is that it is sometimes necessary for the committee to make decisions on points not covered by the rules of the club. In no organization can the committee contradict the rules of the club. These are fixed in the usual way at the Annual General Meeting where, of course, every member has the opportunity to say his piece and where the officers are elected. It is here that the important matters of policy are thrashed out and it is here that any grumbling should be done.

Members' Duties

It is as well for a club member to be taught as a novice that he is expected to help in the running of the club. He can help with the basketing of the birds or in some other manner so that the work is not left to everybody else. If he does this he will find that when he starts claiming his right to be heard on certain things he will get a good hearing. The first thing a club member should do is to get a copy of the club rules from the secretary and read them. In this way, he may be able to avoid being disappointed later by being disqualified or penalized for inadvertently breaking the rules.

Federations and Combines

The club of which a fancier is a member will usually be a member of a federation. The most important function of a federation is to obtain cheaper and better transport for the birds. Some federations award positions and prizes and others are content to act as convoying bodies only. In the same way, in some of the larger centres, federations join together into combines and the larger organizations make sense for the economical use of road transporters. In the sprint races, one federation will be sending enough birds to fill the transporter for the first two or three races. Half way through the season the transporter will be only half full and as the longer races approach this is the time to fly with other federations in a combine. It makes sense to send one full transporter rather than three only one-third full. It's not just the saving of fuel but the saving of convoyers' and drivers' wages.

These federations and combines will also have their rules and these should be read carefully. They may vary quite considerably from the rules covering the club races, particularly if the federation or combine has special marking and clock-setting arrangements. In particular the rules governing hours of darkness and hold-overs should be read with care. There is one final set of rules which must be read and they are probably the most important of the lot; these are the rules of the union.

Sometimes the large or widespread organizations can find themselves with unexpected problems. The British International Championship Club was founded to compete in the International races organized from Belgium. In theory, we race under UK rules, Belgians race under Belgian rules, the Dutch under Dutch rules. In practice the race organizers fix the hours of darkness for all otherwise the national results would not be comparable. More confusingly the split-velocity system is used abroad when the velocities have dropped right down. The split-vel is not permitted in UK races but it could mean the UK velocity is not the same as the International velocity. Fortunately the system is used only for the tail-enders but it can cause confusion to the unwary.

The Unions

Every fancier in Great Britain, in order to race pigeons over the long distances, must belong to one of the unions: the Royal Pigeon Racing Association, the North of England Homing Union, the Scottish Homing Union, the Welsh Homing Union or the Irish Homing Union. These are the main ones but there are others. These unions all have their rules and the information contained in them is of the greatest importance. The Royal Pigeon Racing Association is divided into regions, but these regions, although they have considerable powers, are all bound by the rules of the central body. The Royal Pigeon Racing Association issues a set of standard rules which must form the basis of the rule book of any affiliated club.

The unions have a very important part to play in the sport. One of their jobs is to keep an up-to-date record of every identity ring issued and to forward reports of found birds to their owners. This takes up a great amount of the time and money of the unions, but is by no means their most important job. The second task of the unions is to represent the sport before various official bodies.

An example of the work of the unions is the important shooting case of *Hamps. v. Darby*. The prosecution was successfully conducted through the National Homing Union as the Royal Pigeon Racing Association was then named. This case clarified the protection given to racing pigeons and established the legal point that racing pigeons could not be shot at even if they were actually feeding on crops, unless all other means of scaring them off had failed. This case was contested in the High Courts at some cost but to the considerable advantage of pigeon fanciers. In the same way the unions should and frequently do take the lead in prosecuting pigeon shooters and pigeon trappers.

On another occasion it was claimed by the Irish Ministry of Agriculture that pigeons were carriers of, or susceptible to, foot and mouth disease and fowl pest. Experiments conducted by the International Pigeon Federation had shown quite clearly that this was not true and the Irish Union took this matter up with the Government and succeeded in getting the ban on overseas racing relaxed. Incidentally, the International Pigeon Federation is a combination of worldwide pigeon organizations which encourages mutual co-operation between different countries.

One of the more important occasions when the union represents the sport in general is when application is made to local councils for permission to keep pigeons on council estates. In the old days of pigeon racing there was considerable opposition to the keeping of pigeons. Loft appearances are much better now and opposition is decreasing, but some town councillors still believe that pigeon lofts will be unsightly and undesirable on council estates. Here the unions will send a letter of support for the local request. The periodicals of the sport, particularly *The Racing Pigeon*, do a lot of work in this respect and when the situation is handled well, permission is often granted. Between 1945 and 1955, at the height of opposition, something like 300 councils relaxed their rulings and permitted the keeping of pigeons under certain conditions.

Courts of Appeal

The third important function of the unions is as Courts of Appeal. Any fancier, if he is dissatisfied with a decision given by his club or if he has a complaint against another member, may make an appeal to his union. In the case of the Royal Pigeon Racing Association his appeal would go first to the local region and then to the national council if he were still dissatisfied. The method of making the appeals and the time limits in which an appeal must be made are carefully laid down in the rules and in this way the fancier can go to the highest level if he is dissatisfied with the decision reached in his local organization.

Leading Cases

In many cases, of course, the disputes which go to the highest body are disputes on the interpretation of certain rules. What exactly does a certain wording mean? And what was it intended to mean? These are some of the most difficult points that may arise in appeals. It is a great misfortune to the sport that this question of interpretation of rules has not been tackled resolutely, so that the following is possible, and has actually occurred. In one appeal case a decision was reached supporting one interpretation of a certain rule but this interpretation, however, was not recorded to become a supplementary part of union rules. Three years later another appeal on the identical point came before the same union and a completely different interpretation of the rules was decided upon! This, of course, is a handicap to making this last court of appeal recognized and welcomed by all. Indeed, when decisions such as these are given, some fanciers prefer not to bother with appeals and rather suffer an unjust decision.

Here again there are always a few fanciers who will blame the union. The union does this and the union does that, etc., etc., but the union consists of officials elected by the members and if there is anything to be criticized in the union then the right time for it is at the Annual General Meeting of the union. Here it is a little difficult for each individual member to make his voice heard among the vast numbers from different parts of the country but it can be done. For example, in the Royal Pigeon Racing Association, before a motion can be put before the AGM it must first be passed in the fancier's own club and then be passed at the local region and put forward as that region's recommendation to the AGM. Many fanciers find the administration of regional affairs and union affairs boring, but it is of great importance to the sport and even if they are a little tedious the fancier should make the effort to turn up at the meetings and vote. If he does not turn up at the meetings then every time he has a complaint he should think that if he had turned up, perhaps it would have been different.

Velocity Calculation

Another of the jobs that a fancier should not shirk is working out his own velocities and also his own clock variation. He should be able to do this so that, first, he can check the working of the clock setter and secretary and be certain that all is in order and, second, so that if at some time he is called on to help he is of some use. When I first joined the sport this was all done tediously with pencil and paper. Not even the ballpoint pen was in use then. We progressed first to mechanical calculating machines, like the Odner, and to the pocket electronic calculators and now to the computer program. The trouble is that now the sums have been reduced to tapping a key, the theory may be forgotten.

The working out of velocities is not really difficult as long as it is clearly understood at each step what is going on. The velocity is arrived at quite simply from the distance the bird has flown and the time it takes to do it. It is, of course, the average speed over the distance that is found. The distance is given by the

Official Calculator of the organization and is known to a yard. This distance is converted from miles to yards and then to 60ths of a yard and for this purpose tables are generally used. This distance is divided by the flying time of the bird, converted from hours and minutes to seconds. The result of this ordinary long division sum is the speed of the bird in yards per minute. More accurately, the speed is in 60ths of a yard per second, but this, of course, is the same thing. Nowadays, electronic calculators have become so cheap – the working out of velocities using one is a matter of seconds – that the old pencil and paper methods have been abandoned by almost every club.

Corrected Times

This part of the calculation is straightforward, because the time used for velocity calculation must be the corrected time after allowance for clock variation has been made. Clock variations are always a thorn in the side of the secretary and many ways have been proposed to simplify this calculation, but there is only one accurate way. The three things that are known are:

1 The total time, as shown on the clock, when the clock has been running between setting and checking, known as the long run of the clock.

2 The gain or loss compared with the master timer during that long run.

3 The time between setting the clock and timing-in, known as the short run of the clock.

What has to be found is the gain or loss of the clock during that short run. If we take an imaginary case with round figures, the calculation becomes very much simpler. For example, the long run of a clock is 50 hours, and the short run is 40 hours. When the clock is brought in for checking it is found to have gained 100 seconds. In other words during the long run of 50 hours the clock has gained 100 seconds. From this we can see that in one hour the clock would gain one-fiftieth of 100 seconds or 2 seconds. In 40 hours the clock would gain 40 x 2 seconds (1 minute 20 seconds).

This is the clock variation. What has been done to find it is to divide the total gain of the clock by the long run and then multiply it by the short run. With the figures used it was very easy but unfortunately, as any club secretary knows, the figures are always more difficult than this. We can, however, use the same method. First of all we must turn all times into seconds. This will stop any mixing up of minutes and seconds in the sum. As it does not matter whether we divide by the long run first or last we will do it last so as to be able to work with big figures and not tiny fractions. The first step is to multiply the total gain (or loss) by the short run of the clock. The second stage is to divide by the long run.

$$\text{Variation} = \frac{\text{Total gain x Short run}}{\text{Long run}}$$

This gives us a formula we can always use for finding variations.

Summary of Method

To make this even clearer, here is a summary of the necessary steps followed by an example worked out.

1 Find the long run of the clock and convert to seconds.
2 Find the short run of the clock and convert this also to seconds.
3 Find the total gain or loss of the clock at checking time, again measured in seconds.
4 From these, using the formula above, we can find the clock variation at timing-in.
5 Add or subtract this as necessary to the clock time of timing-in. At this stage must also be added or subtracted the time the clock was set slow or fast. These two together give the corrected time of arrival of the bird.
6 From the time of liberation and the corrected time of arrival find the time in seconds the bird is in the air.
7 Reduce the distance from race point to the loft to 60ths of a yard.

8 Divide this distance by the corrected time of flight to give the corrected velocity.

An Example

Now let us take an example and work it out step by step. First of all we must collect all the facts we know.

- (a) Clocks set 6 p.m. Thursday.
  Clock reading 4 seconds slow.
- (b) Clocks checked 8 p.m. Saturday.
  Clock reading 2 minutes 27 seconds fast.
- (c) Birds liberated 7.30 a.m. Saturday.
- (d) Bird timed-in 6.17.45 p.m. Saturday.
- (e) Distance from race point 507 miles 278 yards.

These are all the things we have to work with. We must find the correct velocity. Let us work this out step by step.

1 Find the long run and convert to seconds. Clock reads at setting 5.59.56 Thursday. Clock reads at checking 8.2.27 Saturday.
   Long run is 50.02.31 or 180,151 seconds.

2 Find the short run. Clock reads at setting 5.59.56 Thursday. Bird timed in 6.17.45 Saturday.
   Short run is 48.17.49 or 173,869 seconds.

3 Total gain or loss at checking time:
      setting 4 slow;
      checking 2.27 fast.
   Total gain 2.31 or 151 seconds.

4 $$\text{Variation} = \frac{\text{Total gain} \times \text{Short run}}{\text{Long run}}$$

$$\frac{151 \times 173{,}869}{180{,}151}$$

```
    173,869 x 151
    173,869
     8,693,450
    17,386,900
```

```
180,151)26,254,219(145.7
    180,151

    823,911
    720,604
    1,033,079
     900,755
    1,323,240
    1,261,057
```

Clock variation is, to the nearest second, 146 seconds.

5   As clock is gaining, deduct this from clock time of timing-in:

```
            6. 17. 45
            02. 26

            6. 15. 19
```

At the same time the 4 seconds that the clock was set slow must be added to give the correct arrival time of 6.15.23 p.m.

6   Find time of flight;
    Liberation 7.30 a.m.

```
            18. 15. 23
             7. 30.
```

Time of flight   10. 45. 23   or 38,723 seconds.

7   Distance of race point 507 miles 278 yards.

```
            53,539,200
            16,680
            53,555,880  60ths.
```

8   Divide by corrected time of flight in seconds:

      38,723)53,555,880(1,383.0
         38,723
         ‾‾‾‾‾‾

         14,832
        116,169
        ‾‾‾‾‾‾

        321,598
        309,784
        ‾‾‾‾‾‾

        118,140
        116,169
        ‾‾‾‾‾‾

         1,971

Velocity 1,383.0 y.p.m. (to the first decimal place).

Approximate Velocities

It is easy to think that all the working involved by clock variations is not worth it but if we work out the velocity in the example above without taking into consideration the clock variation, the velocity obtained is enough in some races to win or lose. If both short run and long run are taken to the nearest hour, in the case of our example 48 hours and 50 hours, then the variation will be 48 fiftieths of the total gain of 151 seconds; in other words, about 145 seconds. This is only one second out from the variation obtained by the proper method but the result might not be so near in another example. The approximate velocity found by using the long and short run in hours is quite a good method but still needs more calculation than is often justified by the last positions in a race. I have known secretaries who have guessed variations without using any method at all especially with the very small variations with quartz clocks. This is all right up to a point and the point is reached when the secretary guesses wrongly.

So far, so as not to complicate things, we have only worked out velocities with fast clocks. There are no differences in using the method with slow clocks except that a loss is always added to the clock time. A point to watch also is that the correct long run of the clock is used. The long run is always the time that the competitor's clock has been running between setting and checking. It is never the actual or correct time as shown on the master timer.

Decimal Places

Many velocities are worked out to three or four places of decimals in order to separate the first and the second in a race. This is all wrong and under no circumstances can three or four places be justified. The old bird average has even been awarded on the third place of decimals, which is even worse. This extreme accuracy in velocities can only be justified if the figures we work from are just as accurate. These figures are time and distance; the time is accurate at the best to the second but what about the distance? Of course an official calculator can make a mistake but more important than this is the information calculators have to work with and their mathematical methods.

The Accuracy of Distances

The information they have is the position of the race point and the position of the loft. The official race point is decided by the Union and is usually a car park, or other parking space. These official race points are decided quite arbitrarily and do not always represent the actual centre of the area of liberation. On occasions the transporters may be moved to points other than the official liberation points.

Second is the loft position: a rebuilt trap on a long loft; changing lofts for those people lucky enough to have more than one; a new loft in a different part of the garden. All these can be sources of inaccuracy, apart from the problem of pin pricks. In the past nearly all fanciers were measured on the 6-inch Ordnance Survey map with a pin prick at the loft position. Now some pin pricks are made by a very fine needle and some seem to be made with a pitchfork. The average size is about a twenty-seventh of an inch. Even if this pin prick is absolutely accurately made the distance covered by the hole of the pin prick on the map is nearly 11 yards (10 m). Nowadays, more and more clubs are buying the 25-inch plans on which to mark lofts and this is obviously a great improvement on the old system. But it does lead to another problem and that is the other inaccuracies that can arise.

The most noticeable error was brought in the original report prepared for the Royal Pigeon Racing Association, or NHU as it was then, on the adoption of the Great Circle System of calculating distances. The report says that this system was adopted because it was the only practical method capable of giving an accuracy of 100–200 ft (30–60 m)! This is the figure contained in the report made by the expert mathematician, Mr Jackson, who undertook the work at the request of the National Homing Union. To demonstrate this point, here is an example given by Mr Jackson. Two competitors are flying 100 miles and 120 miles (160 km and 192 km); they have accurate times of flight recorded and make identical velocities. If both distances are a quarter of a mile too much the 100-mile (160-km) flier will have a higher velocity. If both distances are a quarter-mile too little the 120-mile (192-km) competitor will win.

Dead Heats

Enough has been said to show that distance measurements can never be all that they are hoped to be. Frequently they are no more accurate than the nearest 10 yards (9 m). That being so, how can anyone hope to take velocities to the places of decimals they do and give a fair result? The third place of a decimal is after all a measurement in thousandths of a yard per minute! There is no possible excuse for taking velocities to more than one place of decimals. For mathematicians this means to the nearest place and therefore two places are calculated and the second figure used to correct the first. To show this more clearly here are examples: 1,204.19 is called 1,204.2, 1,204.23 is called 1024.2, with .15 and .25 as the dividing figures. Some club secretaries take their velocities to one place of decimals but only work out the first. By this method 1,204.19 would be given as 1,204.1, an obvious injustice to fancier and pigeon. If there are two birds whose velocities (to the nearest place) are the same they should dead-heat for that position and the prize money be divided. It may be awkward but it is fair.

TEN

Making a Start

The Cost of Starting – Wall-Boxes and Temporary Lofts – The Clock – Buying Birds – Late-Breds – What to Buy – Transfers – Where to Buy – 'Champs or Duds?' – Loft Position – First Drop, Overfly – Fitness – Average Prizes – How Far? – The Need for Patience

People come into the sport from all walks of life and in many ways the most fortunate are those that come in as friends of people who are already racing pigeons. To my mind this is the best way for a fancier to begin. He has the advantage of an experienced fancier's knowledge, which can save him hours and days of wasted time. He has the advantage of experienced advice to prevent him wasting money on worthless pigeons and worthless accessories.

The Cost of Starting

The cost of beginning to race pigeons is quite considerable and is more than a young lad can pay out of his own pocket. It is becoming quite common now for youngsters to have their first introduction to the sport as partners, or assistants, to experienced fanciers. With the older starter the problem is not so bad. It is an ideal hobby for someone who has taken early retirement or has been made redundant at an age when he is unlikely to get a fresh appointment. He is almost certain to have a golden handshake or redundancy money but even so I suggest an older starter follows the same path as a young one. It is only too easy to spend money and unfortunately since big business has moved into the sport there are a few people who would take as much as they could get. Caution before you spend a penny; let that be your watchword.

Let us consider briefly what a fancier needs before he can start racing. Before he buys a bird he must have somewhere to keep it. It is useless to buy valuable pigeons and keep them in sugar boxes, and although a temporary loft can be set up, plans should be made from the beginning for a permanent loft. Pigeons can be kept almost anywhere temporarily. An old shed can be used, but large openings must be cut out and wire netting and traps fitted.

Wall-Boxes and Temporary Lofts

Many fanciers have even used sugar boxes to make their first loft. A large wooden box is fitted with the dowel front and a perch in front of it and hung on a wall, and a couple of birds can be kept in there. My grandfather started with wall-boxes like this and with pigeons bought in the London street market, Club Row. However, the disadvantages of a system like this were so many and so obvious that he soon got a proper loft and real pigeons.

A proper loft can be bought ready-made and a small one will cost anything from £400 upwards. If it is made by the fancier the cost of materials will be from £200 up. Suggested designs and minimum requirements of lofts have already been discussed. Of course the problem may be avoided, or at least postponed, if a fancier buys a good second-hand loft. This need not cost him any more than £150. In addition to a loft the fancier will need tins in which to give his birds food, water and grit. Old tin boxes can be used at the start but as soon as possible proper hoppers with covers to them should be used. There is a great danger of disease if birds soil the water or any of their foods, and a properly built hopper will prevent this.

The Clock

One of the things it may be possible to buy second-hand is a timing clock, which will cost about £100. Unfortunately most cheap second-hand clocks are of models being phased out. There are still many fanciers who swear by the old puncturing clocks but they are a dying breed and it is important to check the make and see if they are still approved and still repairable. Those whose purses are deep enough and want to buy a new clock can get one of the latest models at prices ranging from £400–£500. All the new clocks have quartz movements but some of them need computer print-outs. My strong advice is to consult the club clock setter who should be up to date on the latest models and their suitability in the club. Consulting him can have another advantage because the purchase of a clock can, however, be postponed by the friendliness of a club member since the rules allow a fancier to time-in on another member's clock. The clock setter will know if any fancier has a spare clock to lend and this will ease a major expense. The fancier will also require food, grit and other little items for his loft, but as these can be bought in small quantities the initial outlay can be reduced.

Buying Birds

When it comes to buying birds it is possible to spend any amount according to the resources of each fancier; £500 was once paid back in the 1920s for a pigeon from my grandfather's birds, but there is no need for a beginner to do this! If he is lucky, a fancier will give him a couple of pairs of birds to start up. With even more luck he may find these will turn out winners. Others will have to buy their first birds. In many ways this is the better method, since if the novice fancier purchases them carefully he should start with good birds and if he gives them fair handling he should not lose them.

Some fanciers believe that since many mistakes will be made, beginners should start off with stock that has little real value. I do not think that this is working in the right direction since it will only encourage young fanciers to be careless about the birds and may in fact only cause disappointment when they fail to race successfully. It is better, I believe, for the novice fancier to get reasonably good birds so that he knows he competes as nearly as possible on equal terms with the other members of his club. In that way he will be encouraged to learn more and find out whether it is his birds or he himself that is causing the lack of success.

If the novice buys birds he should buy youngsters. I have discussed the purchase of old birds when considering crossbreeding and much of what has been said there applies to the purchase of young birds. There are, however, some important differences. One of the advantages of starting with young birds is that there is usually no trouble in getting the birds to settle to their new home. They need a few days to get accustomed to the loft but there is no need for the elaborate business necessary with old birds. Another and equally important advantage is that while the number of good old birds available for purchase is limited, there is always a considerable number of youngsters available for purchase during the spring and summer. First round youngsters can usually be purchased in March, or April, but a fancier who is anxious to save money can find it advantageous to purchase the second round (bred May) or late-breds. These are birds bred from the third or later rounds of eggs and are bred from July onwards.

Late-Breds

Late-breds are always cheaper than first round youngsters and are not so eagerly sought after since they cannot be raced that year. This is really an advantage to the young fancier since nearly all beginners want to start their racing careers by sending birds to every race and only too many of them, through inexperience, come to grief. Curiously enough, there are many more late-breds than there used to be. Some like the Scottish fancier I mentioned earlier prefer them to the earlier rounds because they are bred in the better weather and when the parents are under less stress. Widowhood fliers, particularly the long-distance ones, also favour rearing one round or even two of late-breds. It cannot be said too often that late-breds require special handling particularly in regard to their early training. Irrespective of what time of the year they are bought, they should always be trained as youngsters, if only a few miles in the race direction. This would normally be when the young birds are getting to the longer race points, and they will have to be trained separately from the early-bred youngsters. Nevertheless, this early training will stand them in good stead and if they can be trained up to 30 miles (48 km), so much the better.

The yearling year is the time when most late-breds fail. They are, according to their ring numbers, yearlings, but in racing development they are less than youngsters. If they are thought of as youngsters, then not much harm will come to them. The training given to old birds is quite often limited to only two to three tosses but late-bred yearlings should be given as full a training as they would have had as youngsters. If they have flown well round home as youngsters, whether or not they have been trained, they should be well developed. If they have just been rested in the loft as youngsters then their muscles will not have been developed and, although training as a yearling may help them, it cannot build up the pigeon that is deficient physically.

Late-bred yearlings should, then, be trained more rigorously than the original yearlings. This is not so difficult for a novice, who has mostly late-bred yearlings in his loft, but for fanciers with a large number of birds bred at the normal time it may prove a complication that he may not think worthwhile. Nevertheless he should make the effort or be prepared to sacrifice those birds. Late-breds, properly treated, are no worse than their early-bred brothers and sisters and for reasons of price the novice will do well to consider them. Late-breds can also present problems when pairing up starts. All yearlings tend to be wild and difficult to settle in their nestboxes and the late-breds are particularly so. Special care has to be taken to see that no fighting breaks out.

What to Buy

In his selection of which birds to buy, the novice will be guided, as before, into choosing birds bred from those of known racing ability that are likely to inherit the qualities of their parents. He need not have much fear concerning where to purchase. There are a handful of pigeon dealers who spend a lot of money advertising valuable pedigrees and performances of long ago and obviously these are not recommended although some good birds have been produced. There are a few others who have established a name for themselves in racing and who breed large numbers from the birds in order to cash in on their successes. Have nothing to do with them either, for it is unlikely that stock that has been used unmercifully in a breeding programme will give the same satisfaction as more careful breeding.

Choose rather a good fancier who lives near you, who can be visited by you and if possible who has been personally recommended to you. He will normally take an interest in your progress if you buy all your birds from him. Indeed, his advice, particularly in mating, will be invaluable since he has spent many years dealing with the same problems as you will have to deal with.

Transfers

A few words of caution are perhaps necessary to prevent the novice picking up some bad bargains. Do not buy unringed birds and do not buy birds for which you are not offered transfers by the proper Unions. The metal ring on the bird's leg is the bird's identity card and it is registered at the appropriate Union. When that bird is sold it must be transferred to you in the books of that Union in the manner laid down in the rules. It does not matter if you can produce a receipt and certificate of purchase, as far as the Union is concerned and as far as the club is concerned that pigeon is not your property until it has been transferred by the Union. Since the person selling you that pigeon has to sign the transfer form as well, it is also your guarantee that the bird he is selling is his property. If a fancier will not give you a transfer then do not pay over any money at all. When you pay your money, see that you get a transfer form signed by him with the receipt. There is always a temptation to buy an unringed bird when it is offered cheap, but it is a temptation that is best resisted. An unringed bird should never be purchased or kept in the loft unless it is of outstanding value or unless the seller and the history of this bird are well known to the purchaser. It is significant that most of the trade papers will not accept advertisements for unringed birds.

Where to Buy

There is the problem of not only what to buy but where to buy it, but the papers catering for the sport, known to most as the 'fancy press' will usually provide ample particulars for a fancier. The only independent one of importance is *The Racing Pigeon* and in its columns are to be found many advertisements in which good stock is offered. If you make a purchase through these advertisements you will very rarely have cause to complain. A year book published by the same company, called *Squills Year Book*, is perhaps of more importance to the novice starting up since this book has a large number of advertisements by those fanciers who wish to dispose of a few youngsters every year. For more details of these publications you should enquire at the publishers, *The Racing Pigeon.*

Apart from the advertisements in the specialist papers, the only other satisfactory way to get birds is through personal contact with fanciers. If you have become a member of a club you can buy birds from a fellow club member, but this is not really recommended as you will be racing these pigeons against their brothers and sisters. You may do better to buy from a member of a neighbouring club with whom you are acquainted. Pigeons can be bought from other sources: in pet shops and in the larger towns in the street markets, like London's Club Row. A fancier will be advised to steer clear of these, for although many of the sellers in these places are honest, there are unfortunately a large proportion who are not so particular and you will find yourself getting birds that have had their rings cut off or birds for which transfers cannot be given.

A development over the last decade or so has been the the stock farm: the best known is undoubtedly that of Louis Massarella who has spent a fortune on buying many champion pigeons. You can go to his lofts and pay large sums of money to buy the very best from his very best. I do not recommend a beginner to do this. Before spending large sums of money the beginner is far better off gaining more experience. Even more experienced fanciers should pause before shelling out. The huge prices paid for some pigeons owe more to their publicity value than to a proven breeding value. Before the descendants of these birds get down to affordable levels there will be several generations of unraced birds. This is where progeny testing rears its ugly head. It is useless paying out big money unless you can obtain comprehensive information about the breeding success of the parents and the race success of those the same way bred. You are paying out the money and you do have the right to know as much as is available about what you are putting out this money for. Note that this has nothing to do with the bird's pedigree. The pedigree is worthless if the bird can neither breed nor race. This is why progeny testing is so much more important than pedigree.

What the novice really needs are inexpensive birds that can be tested on the road. Louis Massarella and the other stud farms have thought of this and do offer special purchases consisting of a handful of birds of their selection at a much more reasonable price. I must confess that I prefer my old grandfather's strain but if I was starting off from scratch, I would be very tempted to have a try with these. In any case, their catalogues make endless, fascinating reading and the pedigrees are quite instructive. If I have a reservation, it is that they do seem to be cultivating the Belgian strains too much, and I think for the 500-mile (800-km) races British pigeons can still show the Belgians a thing or two.

There are good long-distance Belgian and Dutch birds but they do need sorting out. For sprinters on Widowhood many successful racers believe in Belgian birds bred down from generations of Widowhood sprinters. There is some sense in this provided they are not expected to do anything more than sprint. One of the mysteries to me is why Taiwanese fanciers spend huge sums on international distance winners when because of their geography all their races are sprints!

'Champs or Duds?'

Among the disadvantages of buying young birds is that neither you nor the seller know exactly what is being sold, other than the fact that they look fairly good and do not seem lacking in any important part of their make-up. They may turn out to be champions; they may turn out to be duds; if they seem to be duds you will be well advised to keep on trying with them even when the case seems pretty hopeless. If the birds have been purchased on the lines recommended then the chances of getting a dud are smaller, but the chances are always there. Ordinary care should be taken to see that the birds are in good physical condition when purchased, and if it is not possible to buy them from a fancier who is personally interested in seeing that you get good birds, then you would be well advised to persuade an interested fancier to come with you when you make your purchases.

Another disadvantage of buying youngsters is that they are more susceptible to go down if not handled properly. Nearly all the diseases and illnesses of pigeons occur in their most dangerous form with youngsters and it is particularly important that no chances should be taken with their health. Good food, clean water and plenty of it, a clean loft and plenty of fresh air are vitally important. A word of warning about young birds and strays. Every fancier knows that if a bird arrives at his loft flown out and exhausted, he should look after it until it is fit to continue its journey. One fancier I know did this, but let the youngster mix with his own. The stray was diseased and he had to kill nearly twenty youngsters that were infected. Look after the strays but keep them separate.

When you purchase your birds you should try to get the pedigree of each. If a fancier seems unwilling to give you one, it is not worth your while badgering him for it. A large number of good racing fanciers do not bother with keeping pedigrees except in their heads, and if they are pushed into giving written pedigrees they will have to guess and possibly invent to fill up the sheet. Pedigrees most of the time are not worth the paper they are written on, but a pedigree which lists not only the birds but their best performances is usually worth having. Even so, the methods and position of the fancier must be taken into consideration. A pedigree is a useful thing to have about the place and usually impresses visitors, but unless the fancier himself is acquainted with the birds on that pedigree and unless he can analyse that pedigree to show the system used in building it up, then its value may have been overestimated.

Loft Position

A final point when considering which fancier to purchase birds from remains to be mentioned. It is the position of the fancier's loft. It is undoubtedly true that position can play a part in deciding which fancier will be the winner. This can happen in two ways. Firstly topographical features play a part. For example, pigeons will follow a valley rather than fly straight over a mountain. They will follow the coastline of a bay rather than fly straight across the bay. If flying from the Continent to this country, when they strike the coast they will tend to follow the coast rather than fly at a sharp angle to it. These are generally known matters that cause a lot of discussion and quite frequently the formation of new organizations or special rules to meet the problem. These receive their full measure of discussion, but even more widely discussed are the differences between members of a club. So-and-so always wins the club because he is in a good position; another fancier, usually the one complaining, is in a bad position so never wins. A lot of this talk is just a waste of breath.

Entirely different is the matter of loft position and wind direction. Pigeons do not fly in a straight line from the race point to the loft. Their course is a curve so that if the wind is in the west, they approach the loft out of the east. The French Army and pigeon organizations actually plotted the path of the birds in a race from Morcenz and showed this curved path. Since the birds approach the loft into the wind, it is obvious that the loft on the leeward side will get their birds first and, since the other birds must continue with a head wind, will make higher velocities.

First Drop, Overfly

Fanciers call this advantage 'having first drop'. With a tail wind, the birds are carried over the central lofts and have to work back, hence the lofts to benefit are those having overfly. Having made this clear let it be said that no first drop or overfly made a good pigeon out of bad. Good fit pigeons will always win in the end, although they may lose occasionally when the luck (and the wind) are against them.

Fitness

In the question of fitness lies one of the major causes of a novice's lack of success. It is impossible to write and give the signs of fitness, simply because as in everything else connected with pigeons, it varies with individual pigeons. What must never be forgotten, though, is that unless a fit pigeon is sent, anything else that a fancier has done will be wasted. The right food, the right loft, the right breeding, the right training, the right incentive, all these will be wasted unless the pigeon is fit.

The first thing a fancier must do is to learn to recognize when a pigeon is fit and when it is not quite fit. A sick bird is easy enough to spot. It seems listless, the head is hunched down, the feathers are all ruffled up, the eye is dull and the wattle grey. These signs are easy to spot, but when it is a question not of illness but of being off the peak of condition, then this is much more difficult. Indeed some fanciers will never get to the top because they cannot recognize when their birds are in condition and when they are not. For this reason they send them on days when it would be better to keep them at home, and lose them because the pigeons are given an unfair test.

The best way to discover condition is to go round to the loft of a winning top-class fancier and ask him to show you. Those in the peak of condition will be tight in the feather, firm in the body, clear in the eye and white in the wattle. They will be alert about the loft and will look happy and contented pigeons. As in everything else connected with pigeons, fitness is a thing which a fancier can only discover by constant attention to his own birds. If they are treated properly he will soon learn to recognize when they reach their peak of fitness and will learn when they are just off form. It is worth remembering here that one of the great advantages of the Widowhood system is that the enforced celibacy of the cock keeps it in a much fitter condition for longer periods.

Average Prizes

The greatest enemy of fitness is the average prize, to my mind one of the curses of the fancy. To win an average prize the novice must send and time-in a bird to every race. When the fancier is near the top of the averages and the birds are going to the longer races, there is a terrible temptation to send a bird that is not absolutely fit. The smaller the loft the greater the temptation. What is doubly unfortunate is that these birds are sent to the 400–500-mile (640–800-km) races, some without a chance of getting back, and some that will only get back with luck. The average prizes are a nuisance also in that equal attention is paid to the short races as to the long races. This means that not only must a fancier fly in every race but that whatever he would like to do he must send his birds through to the bitter end at the risk of spoiling his youngsters in the earlier races and of his next year's chances in the later races. The novice will be well advised to leave the average prizes alone for the first year or two. If he tries to keep in the averages by using his yearlings, he runs yet another danger of over-working these not yet fully developed birds.

How Far?

It is on the question of yearlings that there is the greatest amount of disagreement. Some people believe in resting yearlings and only sending them to the shorter races finishing at about 100 miles (160 km). The other extreme is to send them as far as possible, making a 500-mile (800-km) limit. To a certain extent this problem has to be faced in the question of youngsters but only to a lesser degree. Some fanciers say that youngsters must be given the opportunity to grow in peace and therefore should not be raced at all; others say that youngsters, paticularly hens, must be tested and tested severely to know which are worth keeping.

As is usual in most of these cases, the proper answer lies somewhere in the middle. All birds should, if possible, be flown in a race in their first year. With late-breds it is not possible but it is most important that they should be trained down the road. In winter when the fancier leaves for work in the dark, it may be days or even weeks between the times he is able to let his birds out. This will not hurt the birds necessarily, but if at the end of the winter he has birds that have not flown, then he will get a rude shock when he opens his loft on a fine morning and some of the birds he has fed all winter clear off for good.

Another reason for testing young birds is to find out something, even if it is only a little, about the pairings of that year. Every fancier knows only too well that what he hopes will be the best mating of the year sometimes turns out to be a complete failure, the birds bred not being half as successful as he hoped. There is only one way in which he can find out if he has made a successful mating and that is by finding out if the youngsters from that pair will race. Whilst it is true that a good youngster may not be a good old bird, it is also true that if the bird is a complete failure as a youngster it will be a complete failure as an old bird. If he does not race young birds he cannot know if his matings were successful; in other words he is throwing away a whole year of breeding.

At the other end of the scale, he will be severely handicapping his birds if he overworks them. The moult is a time when heavy demands are made upon a young bird. For young birds bred in March or April this is also a time of the longest young bird races. If a fancier persists in racing every week right through to the end, then he may be impairing further growth of these youngsters. This is particularly true, of course, if there have been hard races in the young bird series. This is one of the reasons why YB specialists pair up in December to breed early January and to get the worst of the moult over before YB races, particularly the long ones.

In the same way yearlings are still maturing and must be treated with reasonable care if they are to be outstanding as old birds. This, of course, is particularly true of late-bred yearlings

244

and with them great care must be taken to see that they are not overworked while they are still growing.

The Need for Patience

The question of overworking racing birds is one of the greatest pitfalls in the path of the novice. Normally he will start with young birds and in his second year will have young birds and yearlings only to race. This means that for two years he cannot hope to compete on equal terms with fanciers with established teams. He must be patient and wait until his third year before he attempts to fly a full programme. Patience is indeed perhaps the greatest need of the novice. He must have patience to learn from his mistakes and try again; patience to wait and watch so that he knows his birds well, and patience to endure the bad luck that can indeed face the best of fanciers on occasions.

Index